Freshwater
FISHERIES MANAGEMENT

SECOND EDITION

EDITED BY

ROBIN TEMPLETON

Fishing News Books

Freshwater Fisheries Management

Freshwater Fisheries Management

Editor

Robin G. Templeton

Fishing News Books

Copyright © 1995 National Rivers Authority,
  Severn-Trent Region
Fishing News Books
A division of Blackwell Science Ltd
Editorial Offices:
Osney Mead, Oxford OX2 0EL
25 John Street, London WC1N 2BL
23 Ainslie Place, Edinburgh EH3 6AJ
238 Main Street, Cambridge,
  MA 02142, USA
54 University Street, Carlton,
  Victoria 3053, Australia

Other Editorial Offices:
Arnette Blackwell SA
1, rue de Lille,
75007 Paris, France

Blackwell Wissenschafts-Verlag GmbH
Kurfürstendamm 57
10707 Berlin
Germany

Blackwell MZV
Feldgasse 13
A-1238 Wien
Austria

First published 1995

Set by DP Photosetting, Aylesbury, Bucks
Printed and bound in Great Britain at
  the Alden Press Limited, Oxford and Northampton

DISTRIBUTORS

Marston Book Services Ltd
PO Box 87
Oxford OX2 0DT
(*Orders:* Tel: 01865 206206
     Fax: 01865 721205
     Telex: 83355 MEDBOK G)

Australia
  Blackwell Science Pty Ltd
  54 University Street
  Carlton, Victoria 3053
  (Orders: Tel: (03) 347-5552)

USA
  Blackwell Science, Inc.
  238 Main Street
  Cambridge, MA 02142
  (*Orders:* Tel: 800 215-1000
         617 876-7000
      Fax: 617 492-5263)

Canada
  Oxford University Press
  70 Wynford Drive
  Don Mills
  Ontario M3C 1J9
  (*Orders:* Tel: 416 441 2941)

Library of Congress
Cataloging-in-Publication Data

Freshwater fisheries management/editor, Robin G.
  Templeton.—2nd ed.
      p.   cm.
      Includes bibliographical references (p.   ) and
  index.
      ISBN 0–85238–209–X
      1. Fishery management—British Isles.
  2. Fisheries—British Isles.   I. Templeton, Robin G.
  SH255.F74   1995
  639.3′1′0941—dc20   94–33385
                                                  CIP

A Catalogue record for this book is available from the
British Library

ISBN 0–85238–209–X

Contents

Preface

Since the first edition of this book was published in 1984 there have been several developments and changes in the field of freshwater fisheries management. This second edition reflects these political and technical developments.

With the formation of the National Rivers Authority (NRA) in 1989 (see page 13 and Appendix 3) conservation in all its forms was given a higher profile. Routines and procedures were set up to ensure that the work of the NRA functions (flood defence, pollution control, fisheries, recreation, conservation, navigation and catchment management) at least protected the environment if it could not enhance it. Fisheries management has therefore become much more conservation minded.

All the most common problems are covered, and in those instances where space considerations preclude a full and detailed treatment it is hoped that the updated reading list and appendices will lead the reader quickly to further sources of more detailed or more specialized information.

Part 1 of the book describes the resource over which the fisheries manager holds responsibility. It subdivides naturally into a description of the physical environment inhabited by fish, followed by various aspects of fish biology and fish populations. More information is given on the developments in electric fishing and sonar equipment and their use in surveying fish populations.

Part 2 describes the principal techniques available to the manager for the management and improvement of that resource, be it by direct action such as the draining or desilting of a lake, or by indirect action such as the application of statutory or local regulations. The popularity of building pond fisheries is reflected with more information on the creation and management of stillwater fisheries.

Part 3 examines the commercial exploitation of the resource, with examples of the needs, methods and potential of angling, commercial fishing and fish farming including an expanded Section on the aquaculture of cyprinids.

The 13 appendices are aimed primarily at students of fisheries management in the United Kingdom and Ireland, but again their content will be found to have much wider application and relevance. Descriptions are given of the main organizations responsible for administration of fisheries, and for providing grants to fisheries developments. Information is given on careers in freshwater fisheries and on major employers in the field; and a summary is given of UK legislation relevant to fisheries management. A checklist is presented of the *dos* and *don'ts* for the fisheries manager wanting to create and manage a pond fishery. Finally a description is given of conservation and recreation needs.

Acknowledgements

The task of re-editing this book was undertaken by R.G. Templeton of the Severn–Trent Region of the NRA. Individual chapters and parts of chapters were updated by the following authors: R.G. Templeton, Dr P.E. Bottomley, Dr B. Broughton, K. W. Easton, J. Gregory, V.L. Holt, Dr A. Starkie, Dr R. North, A. Churchward, R. Millichamp, A. Henshaw and M. Moore.

The photographic material for this book was supplied from various sources including M. Ridgway, R.G. Templeton, D. Ford, J.B. Deeker, D. Lippett, M. Pawson, R. North, D. Holdich and B. Broughton. Thanks are due to A. Winstone and D. Williams of the Welsh Region of the NRA.

Figures 3.10, 3.11, 3.12, 3.16 and 3.17 appeared originally in *Salmon fisheries of Scotland*, published by Fishing News Books in 1977 and were prepared by Mrs M. Gammie, illustrator at the Marine Laboratory, Aberdeen. J.M. Templeton prepared Figs 1.23, 1.24, 1.29, 1.30, 1.31, 2.26, 2.29, 2.30 and 3.20.

Whilst the views expressed are those of individual authors who must take the credit for updating their contributions, the production of this book would not have been possible without the support of the Severn–Trent Region of the National Rivers Authority.

The authors

All the contributors are or were employed with the STWA and/or the NRA. They have between them a wealth of experience in all aspects of fisheries management. They have written and published many scientific papers. Most of them are called on at regular intervals, by outside organizations, to give lectures or talks on some aspect of fisheries or conservation management.

Robin Templeton agreed to revise and update the first edition of this book. He was in the water industry from 1965 and spent some time in Hampshire and Yorkshire prior to his final appointment as Area Fisheries Manager in the Trent Area of the NRA, a post he occupied until his untimely death in late 1992. He graduated with a Zoology degree in 1961 and later gained an MSc. He was a Fellow of the Institute of Fisheries Management (FIFM) and was interested in all aspects of fisheries management.

Dr Peter Bottomley was involved in fisheries and pollution work with the STWA and its predecessors during the period 1952–86. He obtained his degree and doctorate prior to 1952. He is a FIFM. Since his retirement in 1986 he has been acting as a part-time consultant in water quality and fisheries.

Dr Bruno Broughton is a zoology graduate who gained his doctorate in the late 1970s. For 11 years he was employed as fisheries scientist and a fisheries officer with the STWA. Since 1988 he has established his own successful fisheries management consultancy business, providing detailed, independent advice on fisheries matters to angling clubs, land owners, local authorities, non-governmental organizations and industry.

Alan Churchward has been in the water industry since 1972. After graduating he obtained an MSc in 1969. He is now Area Fisheries Manager in the Severn area of the NRA. He has a particular interest in the commercial fisheries of the River Severn.

Martin Cooper graduated with a biology degree in 1969. After several years working on trout farms in Scotland and Denmark he joined the water industry as a pollution officer in the Thames area. He joined the STWA in 1973 and is now Area Fisheries Manager in the Trent area of the NRA.

Keith Easton has a biology degree and an MPhil. He has been with the STWA since 1975 as a fish biologist. He is now a fisheries scientist with the NRA based in the Trent area. He is an FIFM and has interests in improvement techniques for stillwater and flowing water fisheries.

John Gregory graduated with a degree in the biological sciences in 1970. After a 5-year period as a marine fisheries officer in the Solomon Islands he joined the Anglian Water Authority in 1975 as a fisheries biologist. He is now an area manager for fisheries, conservation, recreation and navigation with the Welsh Region NRA. He is an FIFM.

Alan Henshaw graduated with a degree in fisheries science. He joined the water industry in 1984. After a period as a warden on a put-and-take-trout fishery he became the fish rearing officer at the Severn-Trent Region of the NRA's Calverton fish farm.

Valerie Holt has been involved in water recreation and conservation for some 20 years. She is an FIFM and is now with the NRA as a conservation and recreation officer. Her particular interests are pond creation, the use of herbicides and general river habitat improvements.

Ron Millichamp joined the water industry, after leaving the Services, in 1953 as a head bailiff in Northumberland. Ron then had various other posts as head bailiff and rivers inspector before becoming a fisheries officer with the Usk River Authority in 1965. Between 1974 and 1982 he worked with the Welsh Water Authority as a fisheries, recreation and conservation officer. He retired in 1982 to start his own consultancy specializing in fishery law and enforcement. He is an FIFM.

Martin Moore has worked in the Thames Region of the NRA since 1975. He has reared many species of British freshwater fish during 10 years involvement with the development and running of two Thames fish farms. One of his many fishery interests has been the development of crayfish farming in the UK.

Dr Rick North obtained his doctorate after graduating with a Zoology degree in 1971. He is a fisheries scientist in the Severn Area of the NRA. His principal interest is in the biology of salmonids.

Dr Alan Starkie has been with the STWA and NRA since 1975. He graduated with a biology degree and later obtained his doctorate. He is a fisheries scientist in the Severn area of the NRA. His principal interest is in the biology of coarse fish.

Part 1
The resource

1.1 The aquatic environment

The aim of fishery managers was elegantly expressed by Richard Seymour in his book, *Fishery management and keepering*, when he stated that, 'knowing and planning the fishery is ... the first important step'. Whatever the ultimate reason for managing a fishery, the manager will find it easier if he has a basic understanding and appreciation of all aspects of the resource he is to manage, starting with the nature and use of water. It is then – and only then – that he can understand how fish have adapted to their environment and how best he can exploit and manage the whole resource.

Water

Water was crucial in the evolution of all living organisms, and biochemical systems are thus well adapted to function in aqueous solution. Moreover, aquatic life requires water as a support system and, in the case of animals such as fish, a medium in which to move and obtain a supply of oxygen. Terrestrial life also requires a regular supply of water to maintain its biochemical processes. Man uses water for drinking, cooking, laundry and bathing; for industrial processes and waste disposal; for cooling water for power-generating stations, and – in dry areas – for irrigation. The demand for water is great: in western Europe, for example, each person may use more than 200 litres of water per day for domestic purposes.

The water cycle

Water falling as rain or snow may travel along several different natural or man-directed paths (Fig. 1.1). It will flow down a river or pass into groundwater, and may be stored in a reservoir, abstracted for domestic purposes, treated at a sewage works, or evaporate at any stage of the cycle. It can also be taken in by plants and used for conversion into carbohydrates by the process of photosynthesis, before being released again during the process of death and decay. It is essential to any fishery that there is a sufficient quantity of the right quality of water. A good fishery manager will appreciate that it is sometimes very diffi-

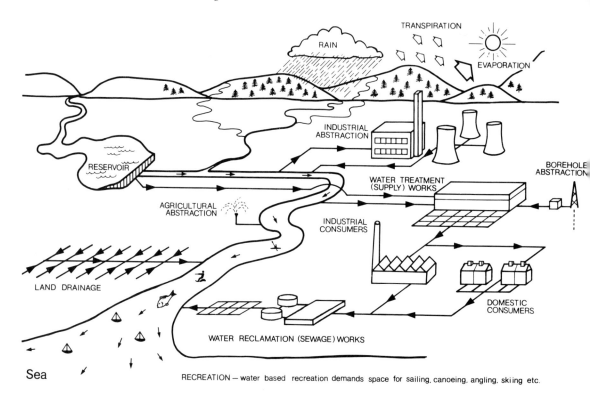

TRANSPIRATION

RAIN

EVAPORATION

INDUSTRIAL
ABSTRACTION

RESERVOIR

BOREHOLE
ABSTRACTION

WATER TREATMENT
(SUPPLY) WORKS

AGRICULTURAL
ABSTRACTION

INDUSTRIAL
CONSUMERS

LAND DRAINAGE

DOMESTIC
CONSUMERS

WATER RECLAMATION (SEWAGE) WORKS

Sea

RECREATION — water based recreation demands space for sailing, canoeing, angling, skiing etc.

Fig. 1.1 The water cycle, showing how water falling as rain is used naturally or by man before being recycled.

cult, if not impossible, to achieve this without a co-ordinated management plan covering an entire river catchment area – of which his fishery may be just a part.

Types of habitat

There is a vast range of freshwater habitats in which fish can live, including mountain streams, slow-flowing lowland rivers, upland tarns, small ponds and large reservoirs. Many books have been written about the different types of freshwater habitat, but the basic division is simply between flowing and still-water.

From the fisheries aspect, flowing water can be sub-divided into four zones, depending on the river-bed gradient and the water velocity. These are described by the principal fish species present. The trout zone (Fig. 1.2) is characterized by the steepest gradient and fastest current; this is then followed by the grayling and barbel zones, and lastly the bream zone (Fig. 1.3) which is very gentle in gradient and slow flowing. In many rivers these zones may overlap or alternate. Fish adapted to one zone may not necessarily survive in

Fig. 1.2 A typical mountain stream characteristic of the trout zone in flowing waters.

Fig. 1.3 A lowland river characteristic of the bream zone of flowing waters.

another because they may not be able to tolerate the physical conditions. For example, the maximum swimming speed of bream is only 0.6 metres per second, whilst barbel can swim at up to 4.4 metres per second. Bream are therefore unlikely to survive in the moderate-flowing barbel reaches of a river where current velocity is in excess of 0.6 metres per second. It obviously makes little sense to stock fish into conditions to which they are not or cannot become adapted.

Stillwaters provide a wide range and variety of living conditions for fish, and the most popular way of categorizing them is based on their biological features. Oligotrophic lakes are essentially nutrient-poor, and consequently never have obvious algal blooms or excessive plant growth; salmonids are the typical fish species present. Eutrophic lakes are nutrient-rich and support algal blooms and abundant plant growth; they may contain a wide variety of the coarse fish species. The term mesotrophic is sometimes used to describe those lakes intermediate in character between oligotrophic and eutrophic.

The total environment

Figure 1.4 shows how fish are related to the aquatic environment in which they live, and it can be seen that their relationships with other animals and plants form a cycle. In its most simple form, chemicals are taken up by plants which are eaten by invertebrates, and these in turn are eaten by fish. When fish die and decompose, their body chemicals are released and are available again for

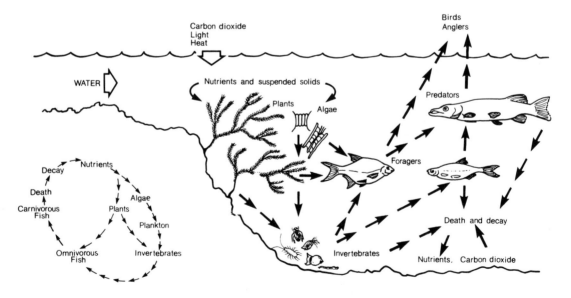

Fig. 1.4 The relationship between fish, plants, animals and other factors in the aquatic environment – 'the aquatic food cycle'.

recycling. It is important to understand this cycle before any fisheries management operations are undertaken because they will, in some way or other, almost certainly alter the environment.

1.2 Water quality

It is a wise fisheries manager who discovers something of the quality of water in or flowing through his fisheries. This is especially important before taking over a new water, because even a brief study of water quality can reveal factors that may have a significant or overriding effect on the number and type of fish that a fishery can support and whether or not it is likely to have reasonably long term benefit for exploitation by angling.

Physical properties

Water has several physical properties that can directly or indirectly affect its quality (Fig. 1.5).

Many of these physical properties, such as its surface tension, evaporation and viscosity, occur because of the way the two hydrogen atoms of the H_2O

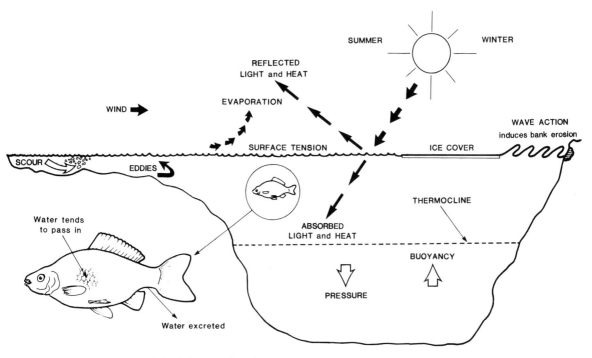

Fig. 1.5 The range of physical properties of water.

water molecule are bound to the oxygen atom. (There is an excellent explanation of this in the *Encyclopaedia Britannica*.) The net effect is that water is able to flow under the influence of gravity from a high altitude to one that is lower, and thence to the sea. One of the direct effects of this water movement is that during its passage it may scour or erode the land over which it flows. The amount of suspended solid material that is collected and carried en route will affect the water's suitability for supporting plant and animal life: at high levels it will increase the turbidity of the water, and therefore deprive plants of light that is essential for growth. When the current slows and the silt or sediment load can no longer be carried, it settles on the river-bed and may smother living organisms that inhabit this area. The gradient of a river and the nature of its bed determines the species of fish most likely to live in it. These 'zones' have been described according to the particle size of the bed material, which is directly related to average current velocity. From the head waters downwards, they are known as 'trout zone', 'grayling zone', 'barbel zone' and 'bream zone' in the most sluggish reaches, having a silty bed.

Water is 700 times denser than air, and aquatic life therefore needs far less body support than land-based life. Water density is also influenced by changes in temperature and pressure. This means that in large and deep stillwaters during the warmer periods of the year, the warmer and less dense water rises to near the surface. Here it is influenced by surface winds and becomes well aerated. In contrast, water becomes most dense at 4°C, and at this temperature it sinks to the lake-bed and gradually becomes deoxygenated. In some large lakes, these zones of warm, less dense water and colder denser water are distinct, and separated from each other by an intermediate layer of water called a thermocline. In winter, when water tends to cool, that which is less than 4°C rises to the surface and a layer of ice may form. It often happens that twice a year the water becomes uniform in temperature and becomes mixed; if the lower layers are deoxygenated at these times, the subsequent 'overturn' can cause fish mortalities. The thermocline, where it is present, varies in depth from year to year and from water to water but it generally occurs at about 10 m below the surface. In some of the newer and larger water supply reservoirs, mechanical means are provided to prevent this thermal stratification occurring: this is not for the benefit of fish but to prevent deoxygenated water from entering the water supply.

Water has the property of absorbing and storing radiant heat from the sun. Rapid overnight and seasonal temperature changes are therefore unusual, except in very shallow and exposed fisheries. Cold-blooded animals (poikilotherms) like fish therefore have a chance to adapt to slowly changing water temperature.

One additional and important property of water is that it will pass through a semi-permeable membrane from an area of low salt concentration into an area of high salt concentration by a process known as osmosis. Freshwater fish have body fluids that are more concentrated than the water in which they swim, and their skin acts as a semi-permeable membrane; they are therefore at risk of being waterlogged. They counteract this by possessing well-developed kidneys to excrete excess water and a protective layer of scales and mucus to help prevent too much water entering their bodies.

Chemical properties

Water is a very good solvent for many different chemical compounds, and without these dissolved substances it would not be a suitable medium for aquatic life. Just as water can scour sediments, so, when it flows down the river channel, it can dissolve the more soluble components of the rocks over which it flows and can leach out soluble chemicals from soil. The surface geology of a catchment is therefore very important when considering the development of a fishery. All aquatic life requires certain minimum concentrations of various chemicals in order to exist. One of the main groups of soluble chemicals comprises the gases, and the two most important ones for aquatic life are oxygen and carbon dioxide. Each dissolves at a different rate in water, and the quantity dissolved is affected by temperature. At 5°C, for instance, water requires 12.7 milligrams per litre (mg/l) of oxygen to become saturated, whereas at 20°C this has decreased to 9.1 mg/l. when the quantity of oxygen exceeds saturation, water is said to be 'supersaturated'. This can happen in summer when abundant plant-life produces vast amounts of oxygen during sunny weather. It should be remembered that different fish have different minimum requirements of dissolved oxygen (DO) below which they will die (Fig. 1.6).

Carbon dioxide is an essential raw material in the process by which plants make their food. This process is called photosynthesis, and is the means by which life on Earth is kept going by capturing the sun's energy. Without this energy input life on this planet would run down and cease to exist. The sun's rays acting on chemicals in green plants enable them to convert carbon dioxide (dissolved in the water in the case of water-plants) and water into sugars and starches. These are subsequently broken down inside the plant to provide the energy for growth and reproduction. During daylight carbon dioxide is removed from the water by green plants and oxygen is released. In darkness, oxygen is absorbed and carbon dioxide is released. Carbon dioxide is released as a by-product of respiration, the process whereby plants or animals convert carbohydrates or sugars into energy, and an end-product of the oxidation processes of 'rotting' carried out by bacteria. This diurnal variation in dissolved oxygen will be discussed within the next few paragraphs; if excessive, it can be a very important factor in the health of a fishery.

The demand for oxygen by bacteria in water is measured and expressed as the Biochemical Oxygen Demand (BOD). It is expressed as the amount of

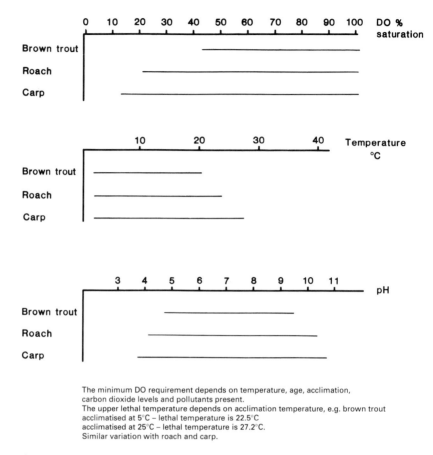

The minimum DO requirement depends on temperature, age, acclimation, carbon dioxide levels and pollutants present.
The upper lethal temperature depends on acclimation temperature, e.g. brown trout acclimatised at 5°C – lethal temperature is 22.5°C acclimatised at 25°C – lethal temperature is 27.2°C. Similar variation with roach and carp.

Fig. 1.6 Approximate dissolved oxygen, temperature and pH range requirements for three species of fish.

oxygen in a sample of water consumed during a period of five days at 20°C, and is generally used to give an indication of the degree of organic pollution in water.

The water molecule in its normal state is split up into two types of electrically charged particles called ions – hydrogen ions (positively charged) and hydroxyl ions (negatively charged) – with both present in equal proportions. If another acid or alkaline base is added to the water, the balance of ions is altered. The degree of acidity or alkalinity is measured on a scale known as pH, which is simply a measure (on an inverse logarithmic scale) of the concentration of hydrogen ions in a given volume of water. A scale ranging from 1 to 14 is used. At a value of 7, water is said to be neutral; below 7 it is acid and above 7 it is alkaline. Most natural waters have a pH within the range 5.5 to 10, the majority being between 7.0 and 9.0. The existence of fish in waters outside this range is rare; sometimes a few stunted trout can be found in peaty streams.

As rain falls through the air it absorbs carbon dioxide and levels may increase further as it percolates through the soil, especially in limestone or chalk areas. It thus becomes slightly acidic due to the carbonic acid formed, and readily dissolves the calcium present in the rock. The hardness of a water is principally a measure of the quantity of calcium bicarbonate $Ca(HCO_3)_2$ dissolved in the water, and levels can range from less than 20 mg/l (soft) to over 300 mg/l (very hard) expressed as $CaCO_3$. Hardness can affect the toxicity of certain metallic ions to fishes. For example, galvanized containers for fish are good when filled with hard spring water, but may cause fish to die if filled with soft moorland water. Aquatic plants need sufficient oxygen, carbon dioxide, a suitable range of water hardness and pH, nitrogen (as nitrate), phosphorus, silicon, potassium, magnesium, iron and other trace elements. In most hard waters these chemicals occur naturally; in eutrophic water there is an excess of certain essential chemicals (usually nitrates and phosphates) leading to an over-production of plant-life. The most common side effect of this is that plants produce excess oxygen as a by-product of photosynthesis in daylight, but at night they absorb oxygen from the water. When plants are very abundant, especially if they are dying and decomposing as well as respiring, they may cause night-time oxygen levels to fall so low that fish become asphyxiated. Under such conditions, plants also cause a night and day change in the pH level.

Waters with few dissolved substances are generally soft, and often drain peaty or insoluble rocky uplands. They support their own fauna and flora, but these are normally impoverished compared with those of nutrient-rich areas.

A measure of the total quantity of dissolved chemicals in a water sample can be obtained by measuring its electrical conductivity expressed as micro-siemens per cubic centimetre ($\mu S/cm^3$), and high figures indicate high concentrations of chemicals. On rivers that receive saline waste water from coal mines, for instance, conductivities may be in excess of 2000 $\mu S/cm^3$; the mountain streams that flow across granite may be conductivities below 100 $\mu S/cm^3$.

A good example of the way in which nutrients are cycled is provided by the nitrogen cycle which begins with the uptake of nitrate for the manufacture of protein. When the plant or animal dies, and fully decomposes, ammonium ions (NH_{4+}) are released and these then become oxidized again to nitrite and eventually to nitrate. A chemical analysis of the amount of ammonia present gives an indication (with BOD) of the degree of organic pollution in a water.

The amount of chloride in a water (its salinity) is also a useful measure in a water sample. Chloride can range from less than 10 mg/l to over 25 000 mg/l (in seawater). Chloride levels increase slightly below sewage discharges, although the levels produced as a result have no harmful effects on freshwater organisms.

Interpretation of water quality data

One of the questions that is most frequently asked of water quality data is 'will this water support more fish?' In order to answer this, and to interpret water quality analysis data, the results are often compared with those of other fisheries.

For convenience it is usual to express the results of a chemical analysis of water quality in the form shown in Table 1.1. The pH is measured on its own scale; most other chemicals are measured by weight, in milligrams per litre. The actual quantities of the various chemicals are thus extremely small. Some substances, such as nitrate (which is always combined with a metal, or with ammonium) are measured by reference to the nitrogen (N) they contain. BOD, the measure of organic activity in the water, is measured by the amount of oxygen the water sample absorbs as a result of this activity under standard laboratory conditions.

Typical water analyses, all from waters suitable for fish to live in, are shown below. The table illustrates the range of chemicals, and measurements expected, from a typical limestone borehole or chalk stream, a large Midland river, and a South Pennine reservoir. (The figures show the extremes where more than one river is sampled, or where more than one sample is taken from the same river.) The chalk stream and the Midland river tend to contain greater quantities of chemicals than the Pennine reservoir. Consequently both produce a greater quantity of plant material and/or plankton for invertebrates, and thus potentially more food for fish. The Midland river carries a larger volume of treated effluent than the limestone water and this is reflected in higher levels of conductivity, ammonia and nitrates. Natural waters that are enriched with effluent in this way tend to provide ideal conditions for coarse fisheries. Conversely, extreme levels of some chemicals, and high tempera-

Table 1.1 Examples of water quality data from three different locations

	Limestone/ chalk stream	Midland river	Pennine reservoir
pH	7.9–8.7	7.2–7.9	6.0–6.9
Temperature (°C)	8.0–15.0	4.0–24.0	4–18
Dissolved oxygen (mg/l)	10–12.6	7.6–12.2	Saturated
Conductivity (micro-siemens)	380–500	640–1090	99–119
BOD (mg/l as O_2)	0.7–4.2	2.0–6.7	1.0–2.8
Chlorides (mg/l as C1)	35–42	50–130	9–14
Nitrates (mg/l as N)	8.5–9.8	5.2–10.3	0.5–1.2
Ammonia (mg/l as N)	0.1–0.5	0.1–0.8	0.001–0.06
Hardness (mg/l as calcium carbonate)	165–210	233–445	32–41
Suspended solids (mg/l)	0–5.0	7–94	2–8

tures, will restrict the all-year-round suitability of such waters for salmon and trout.

It must be remembered that there are seasonal and daily changes in the quantity of chemicals present in any water. In the ideal world it would be useful to have samples taken continuously, but this obviously is not normally possible. Samples are therefore taken at the times of year when conditions are likely to be at their most extreme, i.e. in summer or winter. When only one sample is available, or when samples taken at different times of year are being compared, other evidence (such as the presence or absence of certain invertebrate animal species) should be considered before assessing a water as a fishery. Separate species of fish have their own levels of tolerance to different concentrations of chemicals and also to different temperatures. Figure 1.6 shows the ranges of temperature, pH and DO suitable for trout, roach and carp.

When interpreting water quality data, it is important to remember that an alteration or change in the level of one chemical, or a change in one physical feature, will affect another. An increase in temperature, for example, will usually decrease the level of DO and an increase in temperature and pH will increase the toxicity of ammonia.

Main types of pollution

Most of us think of pollution as being something caused solely by the activities of mankind, but there are many examples of natural pollution – waters with a low pH because of run-off from peaty boglands; waters at the bottom of deep lakes that become deoxygenated; and so on. These, however, fall outside the definition of pollution used here, which is that used by D. Mills in his book *Salmon and Trout*. He considers that,

> 'the natural state of pure water is sometimes upset by man when he allows poisonous substances to enter rivers in order to be diluted and carried away. These substances are in liquid form, and when discharged into the river are known as effluents. Effluents are the waste products of industrial processes or domestic activities and contaminate the water. Such contamination is known as pollution'.

A purist would probably say that pollution is the introduction of any substance or energy (e.g. temperature) into water by man, directly or indirectly that upsets the natural biological balance. At present, the more usual definition is that pollution is the introduction of any substance or energy by man, either directly or indirectly, which would render the water unacceptable for the use or potential use to which it is or may be put.

There are three main categories of pollution, each based on the effect that it has on the environment, the fish and other aquatic life.

The first category covers poisons, including acids and alkalis: the list is long, but includes chromium salts from tanning and electroplating, phenols and cyanides from chemical industries and coal carbonization; copper, lead and zinc from various industries and mines; insecticides from agriculture, forestry, sheep dips and carpet manufacture. It is worth noting that some of these substances are rapidly precipitated in the receiving water (e.g. copper and lead when they run into hard water), but others are more persistent. These poisons usually enter the fish through its gills, and will kill it if present in a sufficiently high concentration. Fish may try, but will not always succeed, in swimming away from poisons.

The second category of pollution includes all suspended solids, the light or finely divided suspended matter which does not settle quickly but makes the water opaque or cloudy. Washing processes associated with mining or quarrying can cause this, as can the washing of root crops. Its immediate effect is to prevent light penetrating the water, thus affecting plant growth. Ultimately such suspended particles settle out and can literally smother and kill all plant and invertebrate life; and if continuously present in large amounts can prevent their growth. This effect can also render gravel unsuitable for trout and salmon spawning by clogging up the spaces between the stones. Where excessive concentrations of suspended solids occur, fishes' gills may be irritated, thus affecting their respiratory process. Protective mucus may also be removed, making it easier for infections of bacteria, fungi and other diseases to enter the body. In extreme cases gill lamellae become blocked and fail to function.

The third category of pollution includes organic residues. Deoxygenation of the water is often caused when these organic materials decompose by the action of bacteria, and the degree of this organic pollution is usually assessed by the BOD test already described. Fortunately, rivers are to some extent self-purifying, the harmful effects of organic pollution usually being oxidized out of the system downstream from the point of entry. The degree and speed of this self-purification is dependent on the quantity of the organic effluent, the size of the river, and how much aeration it receives as it flows downstream. The number of weirs and riffles, for instance, can increase the rate of oxygen uptake in the water and so speed purification. The list of sources of organic waste includes dairies, silage plants, manure heaps, sugar beet factories, textile factories, canneries, breweries, fish meal factories and domestic sewage works. The effect on the river below any point of discharge, particularly if the river is too small to dilute the effluent, is to stimulate the growth of bacteria in the river. These growths in turn lower the oxygen level in the water. This can kill fish by suffocation if the oxygen levels fall too low, or if the fish do not move away from the polluted area.

These then are the three main categories of pollution, but there are others. High water temperatures can make stretches of river unsuitable for certain species (or in extreme cases, all species) and may occur as a result of the waste warm water released by some power stations. Oil is usually visually more dramatic than most other types of pollution: a gallon (4.54 litres) of oil, for example, can spread out over about 2 ha of water surface! There is also a real danger of the water becoming deoxygenated if the oil coverage persists for any amount of time, as contact is lost between water and air. It can also taint the flesh of fish, so this must be borne in mind if the fish caught are going to be eaten.

This section has been necessarily brief. The subject is a large one but it is beyond the scope of this book to go into it in depth. If a fishery is to be successful, however, it must have good quality water (*see reading list*).

Responsible authorities

The NRA was created in 1989 to become the regulatory body for England and Wales for its seven core functions, namely flood defence, water quality, water resources, fisheries, conservation, recreation and navigation (where applicable). The Control of Pollution by Scheduled Processes under the concept of Integrated Pollution Control (set up under the Environmental Protection Act 1990) is carried out by Her Majesty's Inspectorate of Pollution, who are obliged to take advice from the NRA in connection with discharges to water; and that advice must be heeded. The 1989 Water Act privatized the functions previously carried out by the water authorities in regard to the treatment of sewage and the supply of potable water at the same time as setting up the NRA. The Water Resources Act 1991 consolidated the various pieces of legislation relating to the NRA core functions, but much of the Salmon and Freshwater Fisheries Act 1975 still remains on the statute book. The Secretary of State for the Environment has the power to prescribe a classification system for all 'controlled waters' and also to set water quality objectives, to be achieved within a specified timetable. In setting objectives, the minister has to have regard to the EC Directives relating to water quality.

As a matter of routine the NRA takes regular samples of all controlled waters and those discharges (effluents) for which they have control (see also Appendix 1). This continual monitoring helps the authority to ensure that, in the first place, the quality conditions attached to consents to discharge can be controlled and, secondly, that these are having the desired effect on the receiving water. It is vitally important, however, to realise that, no matter what legal powers there are to control pollution, there is always the risk of accidental discharges. If this happens, it may be vital for the first person to notice the incident (often an angler) to report it immediately to the appropriate office of the NRA. In all regions, there is always somebody manning the emergency control rooms who can take the call and put into action the necessary investigations. The telephone numbers of these are set out in Fig. 1.7 and Appendix 3. They are also to be found on a National Angling Licence.

Fig. 1.7 NRA regions in England and Wales.

Head Office
30–34 Albert Embankment
London, SE1 7TL
Tel: (071) 820 0101

North West Region
Richard Fairclough House
Knutsford Road
Warrington WA4 1HG
Tel: (0925) 53999

Welsh Region
Rivers House Plas-yr-Afon
St. Mellons Business Park
St. Mellons
Cardiff CF3 0EG
Tel: (0222) 770 088

Severn Trent Region
Sapphire East
Streetsbrook Road
Solihull
West Midlands B91 1QT
Tel: (021) 711 2324

**Northumbria and
Yorkshire Region**
21 Park Square South
Leeds LS1 2QG
Tel: (0532) 440191

Anglian Region
Kingfisher House
Goldhay Way
Orton Goldhay
Peterborough PE2 0ZR
Tel: (0733) 371 811

South Western Region
Manley House
Kestrel Way
Exeter EX2 7LQ
Tel: (0392) 444 000

Southern Region
Guildbourne House
Chatsworth Road
Worthing
West Sussex BN11 1LD
Tel: (0903) 692

Thames Region
3rd Floor
Kings Meadow House
Kings Meadow Road
Reading RG1 8DQ
Tel: (0734) 535 000

1.3 Water quantity

The water cycle described in Section 1.1 showed how water falling as rain could follow several different paths. The field of science concerned with the natural component of the water cycle is known as hydrology.

Measurement techniques

To aid water management, hydrologists measure catchment rainfall, water losses (to evaporation and to the underground strata) and river flow. River flows in all NRA regions are continuously monitored by purpose-built gauging stations (Fig. 1.8). In the Severn–Trent Region of the NRA (see Fig. 1.7) there are 100 such installations. All major rivers have at least one continuous record of the quantity of flow, and occasional spot flow measurements are also made on an informal basis. Details of all such data are published regularly. Any fisheries manager can obtain an approximate estimate of the speed of river flow if he follows the sequence described in Fig. 1.9.

Water abstractions

Provision for the management of water resources was contained in the Water Resources Act 1963, the objects of which were to promote measures for

Fig. 1.8 A purpose-built gauging station in the Severn-Trent region of the NRA.

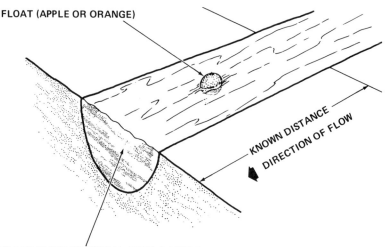

FLOAT (APPLE OR ORANGE)

KNOWN DISTANCE

DIRECTION OF FLOW

AVERAGE CROSS SECTIONAL AREA OF STREAM

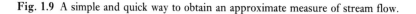

A VERY ROUGH ESTIMATE OF FLOW CAN BE MADE BY MEASURING THE AVERAGE WIDTH AND DEPTH OF WATER (WHERE THE CHANNEL IS REASONABLY UNIFORM) AND NOTING THE TIME IN SECONDS TAKEN BY AN APPLE , ORANGE, OR PIECE OF WOOD TO FLOAT OVER A KNOWN DISTANCE UNDER CONDITIONS OF STILL AIR. THE AVERAGE SPEED OF FLOW OF WATER IN A CHANNEL IS ABOUT 0.8 OF THE SURFACE VELOCITY IN MID-STREAM SO THAT THE SPEED OF FLOW SHOWN BY THE FLOAT SHOULD BE CORRECTED BY MULTIPLYING IT BY 0.8.

THE VOLUME OF FLOWING WATER IS OBTAINED BY MULTIPLYING THE CROSS-SECTIONAL AREA IN SQUARE FEET BY THE (CORRECTED) SPEED OF FLOW IN FEET PER SECOND, WHICH WILL GIVE THE ANSWER IN CUBIC FEET PER SECOND. THIS FIGURE MAY BE CONVERTED TO GALLONS PER SECOND BY MULTIPLYING BY 6¼; TO GALLONS PER MINUTE OR PER HOUR BY MULTIPLYING BY 60 OR 3600.

ALTHOUGH THIS IS A QUICK METHOD, THE RESULTS CAN ONLY BE VERY APPROXIMATE BECAUSE OF THE ERRORS IN TIMING. IT IS IMPORTANT TO TAKE A SERIES OF READINGS AND TO AVERAGE THESE.

Fig. 1.9 A simple and quick way to obtain an approximate measure of stream flow.

conservation, augmentation and proper use of water resources in England and Wales, amongst other things, by the imposition of controls upon the abstraction of water. After various modifications and repeals the present law is now contained in Part II of the Water Resources Act 1991. It is now the general duty of the NRA to take all actions it considers necessary for the purpose of conserving, redistributing or otherwise augmenting water resources and securing the proper use of water resources in England and Wales.

Rivers of good quality with an adequate flow of water are essential for fisheries and other recreational pursuits. Most industries needing water (for example, irrigation water for agriculture and cooling water for power stations), require an abstraction licence. Anyone abstracting less than 5 m³ of water a day (or 20 m³ a day with the NRA's approval) does not require an abstraction licence.

The set procedure for applying for a licence requires the would-be abstractor to submit a copy (to the appropriate regional NRA) of the notice of proposed abstraction and any supporting evidence in a prescribed form, and publish a notice of the proposal in a prescribed form in the *London Gazette* and at least once in each of two successive weeks in one or more newspapers circulating in the relevant locality of the proposed abstraction. Objectors to the proposal have 28 days in which to make written representation to the Authority. If the applicant is dissatisfied with the Authority's decision he may appeal to the Secretary of State. Occupiers of fisheries need therefore to check their local newspapers for applications, or join one of the national angling organizations who monitor all applications in the *London Gazette*. It is worth knowing that the water abstraction licensing section in a regional NRA will usually consult the fisheries section before granting a licence.

The general effect of a licence to abstract water is that the holder is taken to have the right to abstract water to the extent authorized by the licence and in accordance with the conditions contained in it. The holder of a licence may apply to the Authority to revoke or to vary it.

There are a number of exemptions under the Act in that the Authority is prohibited from granting a licence authorizing abstraction or impounding so as to derogate from any rights which, at the time when the application is determined, are 'protected rights'. This means that an NRA Region is in breach of its duty if it grants a licence that derogates from the protected right. The owner who is entitled to that protected right may take court action for damages against the Authority.

Licences to abstract may contain clauses referring to a 'prescribed minimum flow', below which the abstraction should cease. Angling organizations considering purchasing or leasing waters (see Section 3.2) are well advised to check with their local Regional NRA to ensure their proposed fishery is not subject to problems with low flow. Conflict may occur, for example, when a farmer requires water for growing crops or to water cattle – thereby reducing the amount of water to below that necessary for successful fishing.

The NRA has introduced a water resources charging scheme. Details of appropriate charges can be obtained from each of the eight Regions (see Fig. 1.7).

1.4 Basic fish biology

The freshwater fish of the British Isles are of four principal types:

- Fish of the family Salmonidae, e.g. salmon, brown trout and rainbow trout. These are migratory or non-migratory, and are generally cold water fish adapted to fast-flowing water.
- Non-predatory coarse fish, most of which belong to the family Cyprinidae. This family is characterized by the carp and contains the largest number of

species in the British Isles. Its members are all freshwater species and the habitats they colonize vary from upland streams, with clean water and inhabited by barbel, to lowland marshy areas inhabited by bream and tench. In the intermediate habitats live chub, dace, roach and others.

- Predatory coarse fish, of which pike and perch are the most widespread species. This group has been enlarged by the introduction and spread of the zander.
- Non-angled species, an arbitrary grouping composed of small species such as minnow, stone loach, bullhead and stickleback.

There are around 50 species of freshwater fish in Britain and all can be identified with the aid of a suitable key (*see reading list*).

To understand fish biology, a knowledge of the structure of fish is needed. Most European freshwater fish are variations on a similar basic pattern.

As fish are poikilotherms (cold-blooded animals) their body temperature varies in proportion to the temperature of the water surrounding them. Their metabolic rates increase with increasing temperature and decrease with decreasing temperature. Thus any change will control the temperature of any chemical reactions taking place in the fish. This temperature dependence is perhaps the single most important factor influencing fish biology and thus behaviour.

Movement

The external features of a fish are shown in Fig. 1.10. Fish swim by alternately contracting and relaxing the lateral muscles which act on the backbone, causing the tail fin to move from side to side. The fin rays bond as they press against the water and the angle they take forces the fish forward. Different species have tails shaped to suit their speed of swimming: the more the tail approaches the

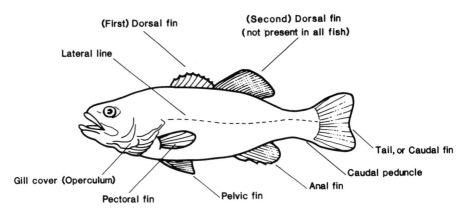

Fig. 1.10 The main external features of a fish.

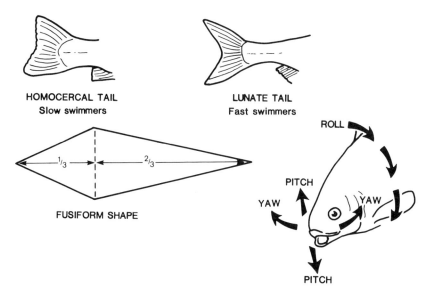

Fig. 1.11 The part tail shape and fins play in swimming and control of movement.

shape of a crescent moon, the faster the fish can swim (see Fig. 1.11). Fins are used for keeping the fish moving straight, and for manoeuvring, and they control the fish in the three planes of pitch, roll and yaw.

Fish species possess different shapes which enable them to adapt to their own particular environment. The faster fish (such as salmon) are streamlined or fusiform in their shape, and the maximum width and depth of their bodies is about one third of the way along from head to tail. There are many variations on this shape, depending upon the environment in which the fish lives. Rudd, for instance, have a shape which is totally different from that of eels.

Skin and colouration

Another feature of the skin is its colour, and this is almost always related to the need for camouflage. Predatory fish like pike and zander have bars of different colour on their flanks to break up their outline and make them more difficult to see by their prey. Surface-living fish such as bleak and dace have silver bellies as this makes them blend into the surface when viewed from below. Bottom-dwelling fish like carp and tench are dark green or brown, the colour of the weeds and the bottom of the lake or stream.

The skin also provides the first defensive barrier against disease and para-sites, and a fish is more susceptible to infection if this barrier is broken (Fig. 1.12). It is very important to remember this when moving fish in and out of their environment for fisheries management purposes. In very cold conditions, when a fish's metabolism is slow, damage inflicted may not be repairable in time to prevent infection becoming serious.

Fig. 1.12 Roach showing damage to scales, through which disease may enter.

Senses

Other external features of a fish include the eyes, the mouth, the nostrils and the lateral line, and these are all connected to the internal organs of the fish (Fig. 1.13).

The eyes of a fish are paired and are located at either side of the front of the head to give them a good field of vision. Fish like pike (Fig. 1.14) have good binocular vision (i.e. depth vision) because their eyes are well forward; tench (Fig. 1.15) have eyes set more on the side of their heads where they are

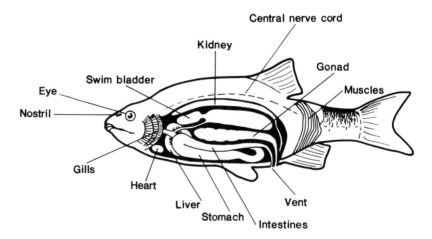

Fig. 1.13 The main internal organs and muscles of a fish.

Fig. 1.14 The location of eyes, well forward on the head, allows good binocular vision, as in this pike.

Fig. 1.15 The location of eyes on the side of the head, as in this tench, allow sensitivity to movement but poorly developed binocular vision.

sensitive to movement but where the binocular vision is poorly developed. The eyes of some fish are bigger and better developed than others – carp, for example, have well developed eyes whereas those of eels and lampreys are small.

The nostrils of fish are paired and normally in front of and just below the eyes. These allow a current of water to enter a sense organ just behind the nostrils. This sense organ is a sac lined with thousands of small cells each of which detect smell. The ability to detect chemicals in solution varies from species to species, although many fish have an acute sense of smell: eels, for example, can detect certain substances at concentrations equivalent to a few drops in the whole of the North Sea.

The lateral line of a fish comprises a band of cells, sensitive to low-pitched vibrations, that normally shows up as a line along the flanks. By using this lateral line, fish are very aware of noise in the environment and can detect vibrations quite a long distance away. For this reason, this lateral line sense is sometimes called 'touch at a distance'. Outboard motors, for instance, produce a large amount of underwater noise which travels much better underwater than it does in the air, and fish have been shown to avoid such sounds even when they are distant. Fish have internal ears but as sound carries a great deal better in water than in air, they have no eardrums or earflaps. Associated with the ears are balancing organs that provide the fish with information on how upright or otherwise it is.

Digestion

The most obvious external feature of a fish is the mouth, with which it breathes and eats. To eat, fish take up food into the mouth, taste it, and swallow it by a suction type of action. All fish in the carp family have throat or pharyngeal teeth which grind the food upwards onto a hard pad so as to crush it before it enters the gut. The design of these teeth is specific to each species and their inspection is one of the only means of distinguishing them. The rest of the gut is concerned with the secretion of digestive juices (enzymes) and the absorption of food. The undigested remains of this food are passed out as faeces from the vent at the rear of the fish.

Respiration

To breathe, fish take water into the mouth with the gill cover (operculum) closed, then close the mouth, open the gill cover, and raise the floor of the mouth. This pumps the water through the gills. Oxygen passes from the water into the blood through the semipermeable membrane of the gill. The blood flow is in the opposite direction to that of the water (called the counter-current principle) so as to obtain the best oxygen uptake. If water flowed in the other direction (i.e. into the gill cover and out through the mouth), the fish might not be able to extract enough oxygen from the water and could drown.

Fish gills are both sensitive and delicate. Air contains 20% oxygen by volume, but water contains only 0.001% at 8°C. Gills, therefore, must be very efficient at extracting oxygen from water, given that there is a sufficient flow of water over them.

Nervous system

The brain is in the middle of the head. It communicates directly to a spinal cord which sends and receives messages to and from the muscles and the rest of the body. The brain of a fish is relatively long, with well developed areas relating to the senses that are important to the fish. The olfactory lobes are often very large as many fish depend on a good sense of smell. The 'higher' brain centres found in mammals are not present in fish so it may be speculated that they live at a lower level of consciousness than warm-blooded vertebrates, and lack rational thought processes.

Reproduction

The reproductive organs of fish are internal. Females carry paired ovaries which rapidly increase in size as the spawning season approaches: just prior to spawning, they may comprise one-fifth of the total body weight. Males have paired testes which do not reach the size ovaries attain. Both the eggs and milt are shed from an opening near the vent into the water, or onto gravel or plants, when the fish are spawning. Fertilization is therefore external. With the exception of tench where the male and the female have differently shaped pelvic fins, it is difficult to distinguish male from female fish using external features alone. In some species size is a good guide (a pike over 5 kg is almost certainly female), whereas in others (e.g. carp) size and sex are not linked.

1.5 Food and food-chains

The processes that comprise the external and internal biology of fishes are often referred to collectively as the 'whole animal response'. It is this that fisheries management attempts to influence, either directly or indirectly. Ecology and the balance of nature in the context of fisheries management are the subject of this section.

What is ecology?

Ecology is the study of the way in which plants and animals relate to their environment. One of the first ecological principles is that plants and animals are a product of their own particular environment and did not, do not, and can never exist apart from that environment. Some of these environmental factors have been described in earlier sections of this book and should be regarded as being indivisible from the welfare of fish.

Food-chains

The relationship between an animal's food and what predates on it is termed a food-chain or food cycle (Fig. 1.16). There are different levels of food producers within this chain, with a limited number of consumers at each so called trophic level. Normally, there is a reduction of biomass (i.e. that mass of animals or plants associated with any particular level) of around 10 to 1 from any one level to the next higher level. This occurs when, say, freshwater

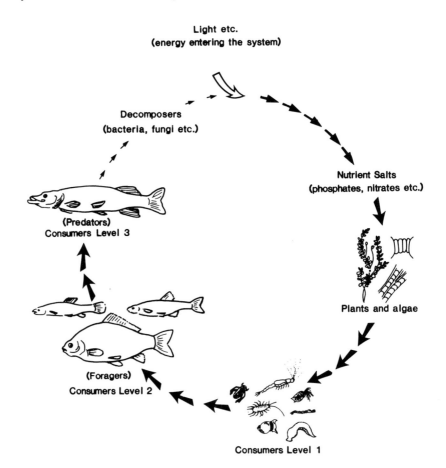

Fig. 1.16 The relationship between different aquatic animals and plants is shown in this 'freshwater food cycle'.

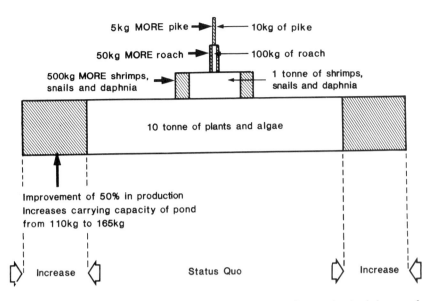

Fig. 1.17 The 'pyramid of numbers': the quantitative relationship between plants and animals in a pond.

shrimps eat plants, or roach eat freshwater shrimps. The best way to understand this food-chain is to consider it as a pyramid of numbers (Fig. 1.17). Thus, any one factor affecting one level will indirectly affect levels further up or down the chain. It follows that predators must, by definition, be far less numerous than the fish on which they feed.

Consider a lake with a healthy fish population of roach and pike. The plants and algae which grow there are eaten at all stages by such aquatic invertebrates as snails, freshwater shrimp and water hoglouse. These in their turn are eaten by the roach. Pike chase and catch the roach, eat and digest them, and grow accordingly. The aim of most pike anglers is to catch a 10 kg fish – and that of the fishery manager is to provide such a fish. This 10 kg pike must have eaten ten times its own weight in roach (100 kg), which together will have eaten 1000 kg of snails, shrimps etc. They in turn, will have eaten 10 t of weed and algae. The energy to form this weed comes from the sun. This whole system is in a state of dynamic balance which is often called the balance of nature, but this term is misleading as it implies a degree of stability that may or may not be present in the system.

There may be little or no actual loss of nutrients in a lake that has neither a feeder stream nor an outflow, because the original materials are recycled many times over. Waste matter excreted by fish and other animals, for example, provides a valuable nutrient source for the plants, and the carbon dioxide given off during respiration contributes to the carbon source. Any organism that dies is broken down by the action of decomposers – the aquatic bacteria which play an important role within the food-chain. (This breaking down process also occurs when animals such as water hog-lice and caddis larvae eat detritus and excrete waste products). It is as important to provide good conditions for decomposers as it is for producers. For instance, the bacteria responsible for making ammonia non-toxic to fish by converting it to nitrate are aerobic, that is, they need oxygen. If the lake or pond bottom is anaerobic and lacks oxygen, very few of these bacteria will survive and this cuts down the recycling of nitrogen in the lake, which in turn may limit plant growth.

As well as the nutrients that are taken into the bodies of animals and plants, light energy is put into the system. This is used to fix atmospheric carbon dioxide and there is also an inflow of nutrients from other outside sources. The source of material may be of great importance in small streams where up to 80% of the nutrients may be obtained from the breakdown of introduced organic matter such as leaves and grass. Thus, the whole balance can be seen to be one of a dynamic flow of energy through different levels in the food-chain.

The factors therefore that govern the number of animals at any level are:

- The number of predators.
- The number of prey (i.e. the number of animals at the level below).
- The number of individuals at each level (i.e. the competition between individuals of the same or different species).

If animals or plants at any particular level are not being eaten they may simply become overcrowded and compete so much for the available food or space that they become stunted. A similar effect may become apparent when predators are removed. Their absence may allow the numbers of animals in the level below to grow unchecked. This is often the consequence of the killing and removal of pike from a small fishery which then becomes over-populated with stunted prey fish.

The role of the habitat

Any one of the levels may get out of control in a lake or pool when, say, the drainage entering it brings with it some of the nitrates used to fertilize the surrounding land. The balance of the levels may have to be restored by removing a certain number of individuals at a particular level, and this concept of 'culling' or 'cropping' is a necessary tool for the management of certain fisheries for specific angler use. This might entail removing weed, culling predators, or reducing the numbers of roach or other coarse fish.

The carrying capacity of a water is of great interest to fisheries managers because the whole of the food-chain is related to water chemistry, sunlight, and nutrients entering the system. There is a well-defined ceiling at which each of the levels in the food-chain can operate, and this effectively limits the number of fish that can be supported there. The sizes and number of fish can vary within certain limits, but it is normally the case that there is a maximum living weight (biomass) of fish that a water can support. If, for instance, a fishery can hold 500 kg of roach, this can comprise 1000×0.5 kg fish or 2000×0.25 kg fish. Manipulation of fish stocks within the carrying capacity is a subject dealt with in the next section. It is possible, though, to improve the habitat so as to increase its overall ability to support aquatic life.

The importance of fish food

It is obvious that to increase the number of fish in a fishery, the whole pyramid of numbers has to be broadened (see shaded sections on Fig. 1.17). It is futile merely to stock it with more fish without increasing the amount of food, because this will merely unbalance the pyramid. The most effective way of increasing the carrying capacity is to broaden the base of the pyramid by increasing the primary production of plants and algae so that this can have a 'knock on effect' upwards. For average lakes in the Midlands of England, a figure of 350 kg of fish per hectare is often the effective maximum biomass of fish that can be supported. In streams, the variation can be much greater.

1.6 Monitoring and controlling fish stocks

One of the key questions asked in relation to river or lake management is: 'How many fish does our water contain?' This is often followed by the questions: 'How many should it hold?' and 'If the right management techniques are carried out, how many could it hold?' It is therefore extremely important to gain a good knowledge of the stock of fish present in a fishery. Once this is known, the techniques required to increase or decrease this stock can be applied. This chapter describes some of the techniques used to capture fish, estimate numbers in a fish population, control stock numbers and keep records.

Fish sampling techniques

There are many techniques available for fish sampling, most of which involve fish capture. For practical purposes they can be grouped as follows:

Seine netting
The most widely used technique for capturing fish is the use of a seine net. This is a simple wall of netting, often 45–90 m long and 3–5 m deep, with corks at the top and lead weights attached to the bottom. It is often fitted with a rope at each end of the cork-line to assist setting. Seine nets are not suitable for most running or heavily weeded waters, and they will only work successfully where the net is deeper than the water (i.e. where the lead-line trails on the bottom).

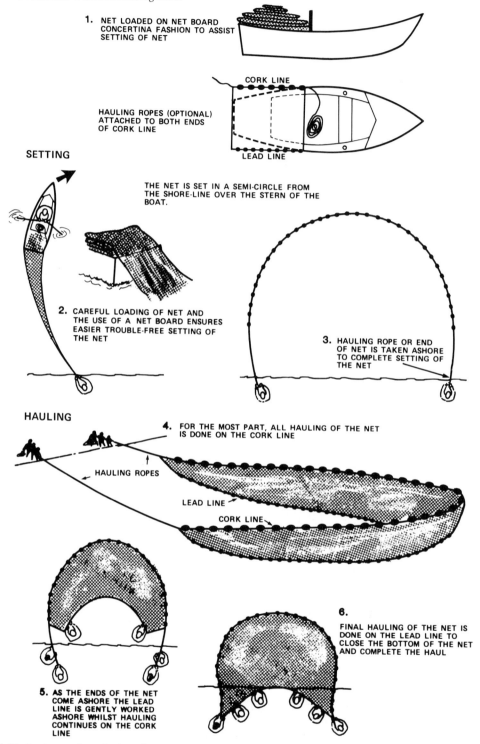

1. NET LOADED ON NET BOARD CONCERTINA FASHION TO ASSIST SETTING OF NET

CORK LINE

HAULING ROPES (OPTIONAL) ATTACHED TO BOTH ENDS OF CORK LINE

LEAD LINE

SETTING

THE NET IS SET IN A SEMI-CIRCLE FROM THE SHORE-LINE OVER THE STERN OF THE BOAT.

2. CAREFUL LOADING OF NET AND THE USE OF A NET BOARD ENSURES EASIER TROUBLE-FREE SETTING OF THE NET

3. HAULING ROPE OR END OF NET IS TAKEN ASHORE TO COMPLETE SETTING OF THE NET

HAULING

4. FOR THE MOST PART, ALL HAULING OF THE NET IS DONE ON THE CORK LINE

HAULING ROPES

LEAD LINE

CORK LINE

5. AS THE ENDS OF THE NET COME ASHORE THE LEAD LINE IS GENTLY WORKED ASHORE WHILST HAULING CONTINUES ON THE CORK LINE

6. FINAL HAULING OF THE NET IS DONE ON THE LEAD LINE TO CLOSE THE BOTTOM OF THE NET AND COMPLETE THE HAUL

Fig. 1.18 A seine net is set by boat from one location on the shoreline and then hauled in.

The net is usually set in a semi-circle from a small boat as shown in Figs 1.18, 1.19 and 1.20. The cork-lines are gently hauled in and slack on the lead-lines is taken up. Too strong a pull on the cork-line will cause the lead-line to

Fig. 1.19 Loading a seine net.

Fig. 1.20 Retrieving a seine net.

rise clear of the bottom. When a small area remains, the lead-lines are pulled in until all are on the bank, thereby trapping the fish in the bag formed by the net. They can then be removed with long-handled landing nets and placed in keep nets or in tubs of aerated water. On long narrow waters, the net can be set across the water and 'walked' down to one end.

Netting problems occur when:

- The net catches on a snag. It can usually be freed by one of the netting team in the boat pulling it upwards from directly above the snag.
- Excessive amounts of silt are dragged in. This can be overcome by 'rocking' the silt-laden net before trying to remove the fish, although this can also damage the trapped fish.
- The water's edge may become boggy. It is wise always to wear thigh waders and select netting sites carefully.
- The net will not 'set' smoothly because of sticks and other debris left in the mesh from previous nettings. They should be removed and the net thoroughly cleaned with water after each netting. It may help if a large sheet of heavy-duty plastic is laid on the bank at the water's edge, so that the net can be placed on this as it is hauled.

It may be useful to carry out a series of seine nettings around a pond or lake. Furthermore, when there is the chance of a large number of fish being frightened out of the netting area, a second seine net can be set and left as a 'stop net' to prevent this. In England and Wales seine netting should only be carried out after permission has been granted from the relevant NRA Region. English Nature must be consulted if any capture methods are to be used in a Site of Special Scientific Interest (SSSI).

It is not unusual these days for some NRA netting teams to survey still waters, prior to netting, with simple sonar equipment. Depth contours can be determined and more often than not the location of fish shoals found (see below).

Electric fishing

Electric fishing can be an excellent means of capturing fish (for qualitative and quantitative sampling, removal of unwanted species, cropping of fish and collection of broodstock for culture), but it is not normally available to angling clubs. The technique is prohibited under the Salmon and Freshwater Fisheries Act 1975, but it can be used under certain circumstances providing prior written permission is obtained from the Regional NRA.

Most electric fishing operations will only catch fish in waters shallower than about 2 m and less than 10 m wide. Usually they are not suited to fish capture in many ponds and lakes, although the use of stop nets may overcome these problems to some extent.

The basis of the technique (see Fig. 1.21) is that the current produced by a

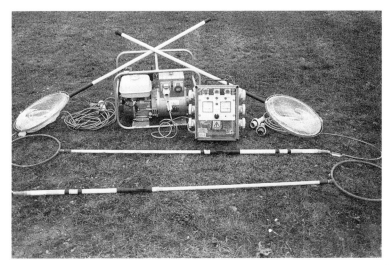

Fig. 1.21 Electric fishing equipment. From front to back: electrodes (pair), control box, generator and hand nets (pair).

generator (or battery) is fed into the water by means of two or more electrodes, thereby creating an electrical field in the water to which fish will respond by some form of forced swimming and/or immobilization, thus rendering themselves easy to capture. Electrical fishing techniques can be sub-divided into two types: alternating current (a.c.) which stuns the fish in its path, and direct current systems (d.c.) which induce fish to swim to the anode. If the d.c. is only allowed to pass in short bursts rather than continuously it is known as pulsed direct current (p.d.c.) and the number of pulses per second is the frequency. In each case, the stunned fish can be removed from the water by hand nets and placed in tanks or keep nets to recover.

Fig. 1.22 Pulsed direct current, boom electrode array in two boats covering greater area in a canal.

A.c. systems are generally used in clear, unobstructed water, whereas d.c. is used in turbid, overgrown or weedy waters. Electric fishing carried out by skilled personnel does not harm fish but prolonged exposure to the electric field, or contact with the electrodes, can cause damage which may prove fatal later. One of the main practical developments of electric fishing equipment has been the use of a p.d.c./d.c. boom electrode array for use only in boats. The typical components of electric fishing equipment and the layout of equipment during field operations is shown in Figs. 1.22 to 1.24.

(a) AC, wading or boat

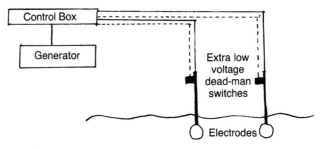

(b) PDC/DC, wading or boat

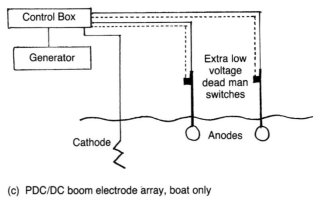

(c) PDC/DC boom electrode array, boat only

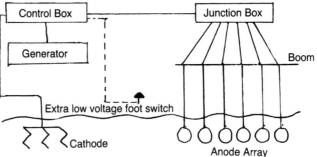

Fig. 1.23 Typical components of electric fishing equipment.

Electrical fishing is also potentially dangerous to the operators and for this reason sophisticated safety devices are incorporated into the system and a strict code of conduct is enforced during all safety operations. Before taking part in any electric fishing operators must pass a medical examination.

The Health and Safety at Work Act 1974 places an obligation on employers to establish for their employees safe systems of work, safe equipment to work with and sufficient information and training to enable them to carry out their

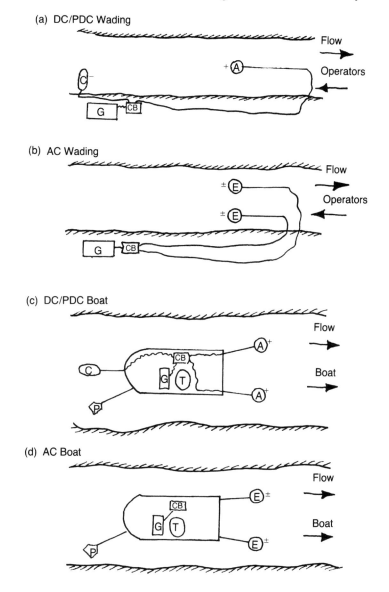

Fig. 1.24 Layout of equipment for electric fishing. (G = generator, CB = control box, T = tank for holding fish, P = paddle, A = anode, C = cathode, E = electrode.)

allocated duties without risk to themselves or others who may be affected by the work being done. Under the Electricity at Work Regulations 1989, electric fishing falls into the category of working near live conductors and, as a consequence, it is a requirement that suitable precautions are taken to prevent injury. In April 1991 the NRA produced its Code of Practice for Safety in Electric Fishing Operations for employees (and those who are working with them or under their control) who carry out, or are associated with electric fishing operations. Although the Electricity at Work Regulations 1989 place a duty on the employer to ensure that every work activity is carried out in such a manner as not to give rise to danger it is essential to note that the employee also has a duty to co-operate with the employer so that safe systems of work are properly implemented. It is recommended that all involved in electric fishing make themselves familiar with the Code of Practice and ensure that they comply with it in full.

Angling

Angling (Fig. 1.25) is often ignored as a sampling technique but can be one of the most useful methods of capturing fish. In particular, angling is one of the

Fig. 1.25 When all other methods fail, samples of fish can often be obtained from anglers.

few methods of adequately sampling large lakes or deep, fast-flowing rivers. Catches may, however, be biased in favour of certain sizes or species of fish, depending upon the techniques used.

The necessary information can be obtained directly from the fish as they are captured, or the fish may be retained in knotless nets until the fishing session is finished. Angling is commonly used in conjunction with mark–recapture studies, described later in Section 1.6.

'Drain-offs'

The removal of water (Fig. 1.26) is an excellent technique for capturing fish but it is rarely practical because it is so expensive. The capital cost of hiring or purchasing pumps is high, and the cost of the fuel they utilize may be considerable.

Pumping can be used to strand fish or to concentrate them into an area or enclosure from which they can then be removed by seine netting or by hand. A total drain-off can yield very precise data on the total numbers and weight of fish (the 'standing crop') present. Local fishery byelaws often stipulate that permission is needed to capture fish legally by the removal of water, and it is wise to consult the local NRA Region before undertaking any drain-off or draw-down work. The effect of drain-offs on any water body is likely to have an environmental impact on all flora and fauna. Invertebrates will survive if some water is retained, but marginal vegetation is likely to suffer if dewatered for several weeks. Lowering water levels will affect breeding birds so any work

Fig. 1.26 The removal of water, drain-off, is a useful technique for capturing fish for examination or counting the total population.

should be avoided in the breeding season (March–July). Great crested newts are protected in law and it is wise to check first for this species before considering drain-offs. Only a licensed person can remove this species.

Echo sounder or sonar surveys

Acoustics cover the whole field of using sound to monitor and detect fish populations. A sonar system that transmits vertically is called an echo sounder; one that transmits horizontally is called a sonar.

The use of echo sounders for fisheries survey work is a technique which has been more widely used in freshwater, since the 1980s, following improvements in equipment design and the production of cheap sounders for sports fishermen (Fig. 1.27).

Sonar was developed for submarine tracking in World War 2, and is an acronym for SOund, Navigation And Ranging. An echo sounder transmits sound pulses and then receives and interprets the returning echoes. When the sound wave transmitted by the echo sounder transducer meets an object such as a fish or the river bed, some of the acoustic energy is reflected back to the transducer. Returning signals are converted into electrical energy and a signal is produced either on the paper trace (Fig. 1.28), the LED screen or the sounder display (depending on the type of equipment in use). A fish or shoal of fish can be determined in terms of its range or distance from the transducer, and the amount of energy reflected gives an indication of the size of the fish or target.

Fig. 1.27 Simple echo sounder, with transducer on right side of boat and sounder with LED screen.

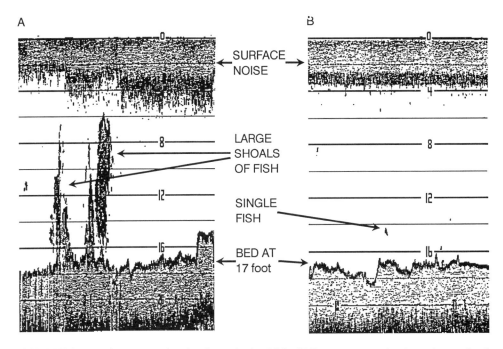

Fig. 1.28 (a) Echo sounder transect showing dense shoals of fish; (b) Repeat transect showing only occasional fish after the shoals had been removed by netting.

The more expensive scientific sounders (Figs. 1.29, 1.30 and 1.31) can be calibrated to produce more accurate and ranging sizing of fish. Computer processing of the accumulated data is also possible to arrive at a figure giving the weight of fish per unit area of water surveyed. Echo sounders can be used to locate fish which are to be captured by other techniques and/or can be used to complement (or in the place of) other survey techniques. It may be possible, for example, to electric fish the weedy marginal zones of lakes or rivers successfully but not the deeper offshore areas. If the bed of these offshore areas is unsuitable for trawling or seine netting and the depth insufficient for mid-water trawling, echo sounding might be a useful and cost effective alternative technique. It is also a useful technique for surveying large areas of water in which fish have gathered in large shoals.

Fixed transducers may also be used to detect (and size) fish moving past a given point and can, for example, be used to determine the behaviour and movement of fish at dam or weir sites. In situations such as this transducers are often mounted 'looking up' from the bed or horizontally out into the water body.

As with other survey techniques echo sounding has its advantages and disadvantages. It is therefore important to determine what information is needed and whether conditions are suitable for its use before embarking on a sonar or indeed any other type of survey. Shallow weedy waters are generally

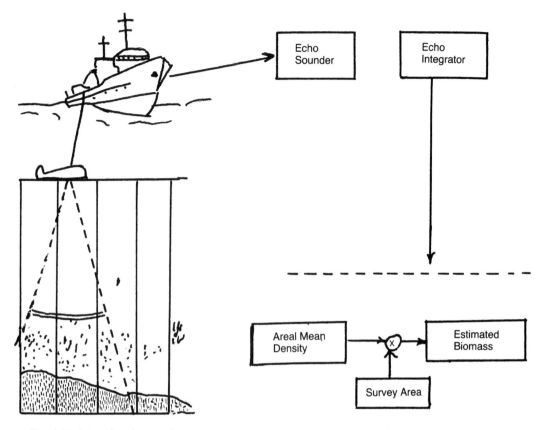

Fig. 1.29 Scientific echo sounder survey.

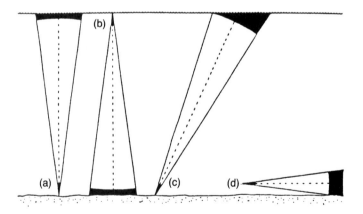

Fig. 1.30 Transducer configurations. Typical transducer aiming configurations include (a) bottom-mounted, aimed straight up; (b) surface-mounted, aimed straight down; (c) bottom-mounted, aimed obliquely up; and (d) bottom-mounted, aimed parallel to the surface. Shaded areas show non-detection zones or biased detection zones.

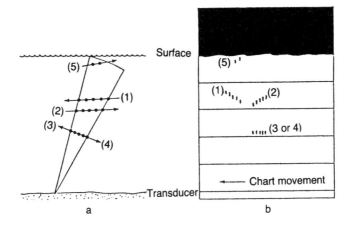

Fish movements in an acoustic beam

Fig. 1.31 Fish movements in an acoustic beam. The general direction of fish movements can be determined from 'trace types' on chart recordings if the transducer is aimed at a non-perpendicular angle to fish movement. In this example, fish movements through an acoustic beam (a) are shown with black dots representing successive acoustic detections. The resulting echo traces are shown in chart recording (b).

unsuitable for echo sounder surveys. Further details of this technique and its use can be found in the *reading list*.

Fish counters

Automatic electronic fish counters are used to count the up and downstream movement of fish, especially migrating salmonids. These counters need to be located at permanent sites such as weirs or fish passes.

Fish moving over 'resistivity' counting strips on a weir create a change in the electronic balance, causing a counter to register the movement. Fish counts obtained in this way are vital to the management of exploitation. Also such counters can be used to determine the success or failure of schemes to reintroduce salmonids into a catchment. To obtain reliable data attention needs to be given to the proper calibration of the counting system (see *reading list*).

Other techniques

There are several other techniques (see Fig. 1.32) which may be used to obtain fish samples, although most require either expensive equipment or are impractical in many fisheries. The nature of the fishery and the type and extent of data required will determine which technique is most suitable, but it would be prudent to seek specialist advice before sampling begins.

Gill nets rely on the fact that fish will swim into a wall of fine net which is either suspended from the water surface or anchored to the bed of the fishery. The mesh size has a major influence in determining the size of fish caught, and

Fig. 1.32 Various nets are available to sample and/or hold fish samples (clockwise: seine net, beam trawl, fyke net, 'Windermere' perch trap).

most sampling will involve using 'gangs' of different sizes of gill nets. Fish mortality rates in gill nets are generally high, and they are therefore often used only to remove unwanted species (e.g. pike) from a fishery. They may also catch species which it is desired to retain. Gill nets are rarely employed in running water.

Trapping is a long-established fish capture technique and many styles of trap exist. Most traps include some form of leader which encourages fish encountering it to enter a central holding chamber. Permanently installed traps include those for migratory fish, such as salmon, and traps can be designed around man-made fish deflectors such as the weirs used for capturing eels. Portable traps include hoop and fyke nets, and 'Windermere' perch traps. Fyke nets need to be fitted with otter guards.

Trawling is essentially a marine technique that has been scaled down and adapted for sampling large freshwaters. The mouth of an otter trawl is kept open by floats, weights and otter boards which act like underwater kites, whereas beam trawls are built on a rigid frame. They may be fished on the bottom or in midwater, and they are often used in conjunction with echo sounder or sonar equipment. Trawls have become widely used in some areas for sampling young fish in large lakes and reservoirs.

Nowadays poisoning is rarely used to achieve a total fish kill prior to fishery renovation. At sub-lethal levels, it can also be a successful method of sampling fish, although considerable care is needed in applying the correct dose of poison, and speed is essential when removing the narcotized fish. The most popular poison is rotenone, a natural plant derivative, and there is considerable scope for the development of less expensive and more easily applied alternatives, such as antimycin-A. Poisons are rarely specific in their action and may act on animals other than fish (including man!). As they are a prohibited fish capture technique under the Salmon and Freshwater Fisheries Act 1975, they must never be considered for use without first consulting the relevant NRA Region. Indeed, special consent of the Ministry of Agriculture, Fisheries and Food must be obtained by the NRA Region before they can be used by anybody.

Population estimates

Following the capture of fish, it is often possible to make some estimate or calculation of the total size of the fish population. One of the most useful measures that can be made is of the weight of fish that a unit area of water contains. This is called the standing crop. Many waters hold 250–400 kg of fish per hectare, although some will support less than this amount. If suitable habitat improvement techniques are applied, it may be possible to elevate this figure to 500, or even to 650 kg/ha. There are three main techniques for estimating population numbers:

Direct counting
Where all the fish have been captured from a water, for instance following a 'drain-off' (Fig. 1.26) it is a simple matter to count and weigh the fish.

Depletion technique
Where data are available from two or more successive samplings it is possible to estimate the total population size. This is only true if, (a) the same technique is applied to the same area each time, and (b) the catch on each occasion is smaller than that of the previous occasion.

If two successive samplings have been carried out,

and C_1 is the number of fish caught in Catch 1
and C_2 is the number of fish caught in Catch 2

then the population estimate (N) is calculated as

$$N = C_1^2/C_1 - C_2$$

This is the simplest type of depletion equation; there are many more complicated versions. The weight of fish present can be calculated by multiplying the estimated number by the average weight of the fish caught.

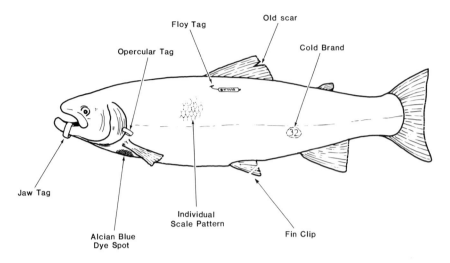

Fig. 1.33 Several methods are available for identifying individual fish.

Mark–recapture techniques

The basis of any mark– recapture technique is that a known number of fish are taken, marked in some way and released. A subsequent sample of fish from the fishery should, if large enough, include some marked and some unmarked fish. The calculation of the population size can be made using various equations, one of the simplest of which is:

$$N = (M \times C)/R$$

Where N = population estimate
      M = number of marked fish in population
      C = total number of fish caught in second sample
      R = number of marked fish recaptured in second sample

The ability to recognize fish because of individual characteristics, tags or marks is most useful. With experience, many individual fish can be identified by their colour and scale pattern, or because of deformities and scars. However, tagging or marking techniques give a more reliable and easily recognized method of identification. Several different methods of tagging and marking are available (Fig. 1.33). Modern technology has developed radio tags. These are usually used for some specific research project (*see reading list*).

Marking

Of the various marking techniques used, the injection of 'alcian blue' dye is the most commonly used and the most successful. A solution of the dye is injected into the skin beneath a scale (or scales) by using either a syringe (usually a 1 ml

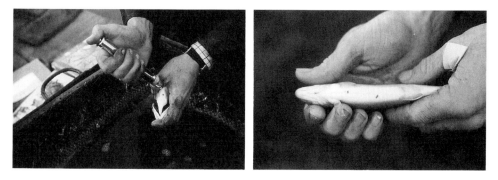

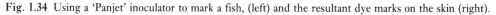

Fig. 1.34 Using a 'Panjet' inoculator to mark a fish, (left) and the resultant dye marks on the skin (right).

or 2 ml disposable plastic syringe with a 23G or 25G needle) or a 'Panjet' inoculator. This instrument propels into the fish, at high pressure, a small volume of the liquid held within its reservoir. It is a dental instrument which requires minor modification by the manufacturer to enable it to be used for fish marking. Although expensive, Panjets are both fast and safe in use (see Fig. 1.34).

The alcian blue can be injected to mark and identify batches of fish by using a common marking position on their bodies such as between the pectoral fins. Individual fish recognition can be achieved by using a combination of a series of marking sites. Alcian blue marks fade, however, and are generally lost after one or two years, although records exist of fish that have retained marks for seven years and more.

Tags

Many different types of tags are used in fisheries biology, and each is suited to a particular need or species. Most are metal or plastic discs or tubes that are wired onto the gill covers or fin roots. They may bear a numbered code which enables the identification of individual fish.

Fin clips

The clipping or cutting away of part of a fin or fins is a popular method of marking fish. Cut fins will eventually regenerate, but evidence of the cut usually remains. Clipping sometimes affects the mobility of fish, and is best restricted to the paired fins or the adipose fin of salmonids.

Stock control

When population estimate figures have been calculated, additions or removals to the stock of fish can be considered.

Fish stocking

It should be stressed that fish stocking should only be undertaken after other means of increasing the standing crop (e.g. habitat improvement) have been considered. In many situations a great deal of money has been wasted by introducing fish to fisheries where their chances of survival were poor or where there was no need to supplement the existing natural stock of fish. Stocking is usually only necessary when there is a deficiency in natural reproduction; where the species has been reduced or eliminated; or where the species has not existed before and where conditions are suitable for that species.

The stocking of fish should be undertaken in a carefully planned manner. The type of species used will depend on the nature of the water to be stocked and the type of fish required. It has been shown that stillwater fish may undertake a high degree of migration when stocked into rivers, thus providing very limited benefit to catches. Where possible, running waters should be stocked with fish that have originated from running water. Fish for re-stocking can be transferred from other waters, purchased from fish farmers, or sometimes obtained from the NRA. Even if it was originally devoid of fish, there is rarely justification for stocking a water with more than 300 kg/ha of fish.

Fish introductions during the summer months are inadvisable. Because of higher water temperatures, fish are less easy to transport and become more stressed than during the cooler months. In Britain, fish stocking after September and before May could be undertaken without these problems. If fish are stocked in spring, they will be able to take advantage of the rapid increase in the numbers of invertebrate animals at this time of year; in autumn, the 'larder' of natural food is diminishing and their survival may be impaired as a result.

Before restocking occurs, it is necessary to obtain prior written permission from the NRA (and from English Nature if the water is an SSSI) who often require a sample of the fish to ensure that they are not heavily diseased or parasitized.

Fish removal

The removal of unwanted or surplus fish can be achieved by using the capture techniques described earlier. In some cases it is possible to sell these fish to commercial fish suppliers, and the funds so raised can be used for other fishery development purposes. If it is not possible or practical to remove fish yourself, most fish suppliers will undertake this service although they will obviously pay a lower price for the fish so obtained. Alternatively, the NRA's Fisheries Department sometimes considers removing unwanted fish.

Keeping records

The compiling and retention of an accurate set of records is an important aspect of good fisheries management. The fishery manager should keep detailed information on stocking dates, species and numbers of fish caught, etc.

If angling competitions are held on the fishery, a complete record of the match results should be retained. Numbers of each species caught are most useful, in addition to total weights obtained by each competitor. This will enable past performance to be compared with present, and the data will give a good indication of any decline or improvement in the fishery. Photographic prints and slides are a means of recording physical features and provide useful additional information to that obtained from maps and depth charts.

Any information that is collected should be stored in an orderly, readily-accessible form and should always include the date on which the information was collected. It does make sense for any fisheries manager or angling club to invest in a personal computer (PC). Records can then be stored safely, and fairly complex analysis undertaken simply and swiftly.

1.7 Fish mortality

Fish, like any other living organism, are not immortal. To the angler, dead fish are often a cause for concern, but it must be remembered that in order to recycle the nutrients within them, the process of death and decomposition is essential for new life.

The premature death of fish can, of course, destroy a fishery, and the appearance of dead fish is often the first obvious sign of a polluting discharge that can threaten other animals or water supplies. The principal causes of mortality are summarized in Fig. 1.35.

Causes attributable to man

Pollution
The main types of pollution have been described earlier. Generally, pollution only causes a fish mortality if it is sudden. An example is the sudden influx of a quantity of silage or piggery waste into a clean watercourse. If the pollution becomes continuous, fish may be eradicated, become accustomed to it, or migrate to unaffected areas, depending upon the volume, its strength and the extent of dilution afforded by the stream.

Natural causes
The enrichment of some still waters causes blooms of blue–green algae. Such algae can cause fish mortalities not only by the excretion of a toxin but also by the very quantity of algae blocking the gills of the fish.

Angler-caused damage
Angling practices can, to a lesser or greater extent, physically damage fish.

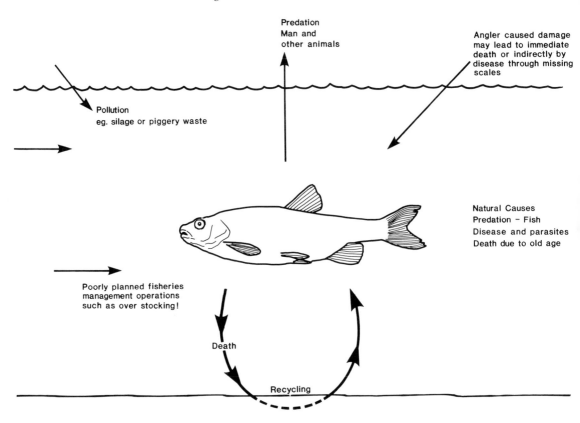

Predation
Man and
other animals

Angler caused damage
may lead to immediate
death or indirectly by
disease through missing
scales

Pollution
eg. silage or piggery waste

Natural Causes
Predation – Fish
Disease and parasites
Death due to old age

Poorly planned fisheries
management operations
such as over stocking!

Death

Recycling

Fig. 1.35 Several causes of fish death.

Hook damage to the mouth is a common injury, but foul-hooking can cause injuries almost anywhere on the body. Minor damage may be limited to the loss of a few scales, but this can nevertheless open the way for organisms like fungi and bacteria to gain entry to the fish.

The retention of fish in keep nets sometimes causes subcutaneous hae-morrhages, split fins, and loss of protective mucus. The damage may be so severe that fish die as a result, or become infected with lethal diseases or parasites. Research programmes may give some indication of the best design of keep net.

Catching fish by any method 'stresses' the fish. Such stress can alter the internal balance of body chemicals and make the fish more susceptible to disease. Stress cannot be avoided but it can, with careful and thoughtful angling practices, be kept to a minimum.

Fisheries management
Poorly-planned fisheries management operations can cause fish mortalities. There are many examples of how well-stocked and balanced fish populations

have been altered by an unnecessary restocking operation, the extra stock causing stress, starvation and a decline in the fishery with the end result being poor fishing.

Natural causes

There are three main natural causes of death in fish: predation, diseases and/or parasites, and old age. It should be borne in mind that fish compensate for these losses by producing far more eggs than could normally survive as adults, and that death from these causes is the method by which fish populations are naturally controlled.

Predation

Predation occurs at all stages of the life-cycle. The eggs and alevins of many fish are eaten by invertebrates like caddis-fly and beetle larvae, and by vertebrate animals such as frogs, birds, various mammals and adult fish. It has been calculated that a 97% mortality in roach from egg to fry was attributable to invertebrate predators, and that of 1.5 million pike eggs in a river population, only 68 fish survived to become large enough to be caught on rod and line.

Many coarse fish fry will also die if the right food is not available at the correct period in their development. The fish which survive these young stages are still subject to predation from other fish and animals, but the older and larger they become, the lower the risk from predation.

Disease and parasites

Fish diseases usually act in conjunction with other environmental factors: stress is particularly important in causing outbreaks. This can happen with a change in water quality which lasts for a considerable time but which is not in itself sufficient to kill fish. It may be stressful enough, however, to make the fish more susceptible to resident disease organisms. Hence, significant fluctuations in the levels of DO, temperature, pH, ammonia, salinity, suspended solids and others may all cause stressful conditions.

Many diseases and parasites are common on fish farms. Poor husbandry techniques can soon lead to severe outbreaks which, if left untreated, can cause large mortalities. Most fish farmers are familiar with the basic diagnostic and treatment techniques, but the best cure for disease on the fish farm is to prevent them occurring in the first place by ensuring that good husbandry techniques are practised.

Examples of large mortalities in the wild are fortunately not common, and most fish will carry a complement of parasites without any untoward effect. There are examples, however, of infestations that have killed or damaged individual fish or whole populations, and they commonly occur in over-crowded fisheries.

Tapeworms like Ligula sp. take food from their host and they may also render it infertile. Other tapeworms block the gut and cause fish to die through a lack of nutrients. Skin parasites can irritate the fish and cause considerable damage to the skin. Wounds caused by parasites often allow the entry of secondary bacterial and fungal infections, and these can frequently lead to the death of the fish. The fish louse, for example, often causes severe skin damage which can become infected, and coupled with the toxin the parasite injects, can prove fatal.

Despite all these hazards it is possible that some fish do, in fact, die of old age, although the exact meaning of this term is the subject of much speculation.

Poor condition

Many anglers are able to recognize when a fish is in good or poor condition (Fig. 1.36). It is possible to calculate the relative 'fitness' of a fish by relating its weight to its length, and this condition factor gives a clue to the healthiness or otherwise of the individual or population.

Most fish populations contain the occasional fish in poor condition, and such specimens are usually found to carry a higher than normal parasite load, or be physically damaged and so unable to feed properly. However, reports of large numbers of poorly conditioned fish from the same water should always be

Fig. 1.36 Trout in poor condition.

investigated as they are usually symptomatic of one of many fisheries management problems.

Poor flesh taste

Complaints are occasionally received that fish taste unpleasant when eaten. It is difficult to evaluate the taste of fish flesh objectively. Towner Coston *et al.* in their book *River Management* in 1936 said that, 'flavour is very largely a question of personal opinion, e.g. some people declare that River Test trout are uneatable, others that they are as good as any other trout.' The flesh of any animal is often only as good as the food that the animal has eaten, and it is possible that a trout can take on the flavour of the invertebrates it has been eating. If those invertebrates have themselves consumed certain types of plankton, then this may result in a poor taste. There appears to be very little the fisheries manager can do about this. Sometimes, however, fish flesh can become tainted due to very small quantities of chemicals such as phenols, which have contaminated the water.

Part 2
Management techniques and methods

Having reviewed the resource that the fishery manager is expected to manage, we shall now consider some of the techniques available to him.

A decision will have to be made at a fairly early stage as to the objectives of management actions and efforts should be made to determine whether or not it is possible to attain these objectives. It is pointless, for instance, deciding that a fishery should be managed for large carp if anglers do not want to fish for them, or do not have the necessary expertise to catch them. If this is the case a good general mixed fishery should be created, as described below.

Natural fisheries by definition are not managed. The species present may only be known from past catches, and no restocking, destocking or habitat improvement operations are undertaken. The facilities may, however, have been improved for the angler. Examples include most of the mixed coarse fisheries of the River Trent and the River Thames, and many salmon rivers. Specialist fisheries are those in which an angler can expect to catch a certain target species. These fisheries are intensively managed, nearly always stocked, and have been subject to habitat improvement works. Examples include put-and-take salmonid fisheries and specialist carp fisheries. Somewhere between these two lie all the other fisheries.

Choice of fishery, coarse of salmonid

The fisheries manager has the choice of creating or restoring a coarse fishery (still or flowing water) or a salmonid fishery (still or flowing water). Mixed fisheries are possible, but are not easy to manage well. There are a number of factors that should be taken into account when considering the construction of a pond fishery, whether or not the pond is in existence or still to be created. It should:

- Support fish life. Duck flighting pools may be too small and shallow; irrigation reservoirs with varying water levels are unsuitable for fish.
- Allow fishermen access to the water and the fishing without endangering them or other members of the public.
- Provide water of a sufficient quality. Trout will not breed in ponds. They prefer a dissolved oxygen (DO) level of 10 to 16 mg/litre which is usually

found when the BOD (see Section 1.2, page 7) is less than 4 mg/litre. Preferred temperatures for trout are 14 to 16°C; at 18 to 20°C they are disturbed and at 25°C they die. Levels of free ammonia should be low. At the other extreme carp, for example, can cope quite well with a DO level less than 5 mg/litre, temperatures up to 35°C and higher levels of free ammonia. Other species such as roach, rudd and perch require DO levels which fall somewhere between those required by trout and carp (see page 8).

- Offer a secure site. Trout and large carp are more likely to be stolen than other species. If the site is not secure, then other species should be selected for stocking.

Other factors to be considered are:

- Costs of stocking/restocking. Trout fisheries, which are dealt with in Section 2.2, are usually stocked at about 100 kg/ha. Because trout do not breed in ponds there has to be regular restocking to replace fish which are caught and removed. Coarse fisheries are more expensive to stock initially, but don't usually require further stocking, and indeed may in later years produce excess stocks.
- Likely income/profitability. It is difficult to be accurate about income as it very much depends on supply and demand. Rental fees are often higher for trout lakes. The cost of a day ticket on a trout fishery is almost always higher than on a coarse fishery, because an allowance has to be made for the value of fish removed. Trout anglers generally expect to take their fish home whereas coarse fish anglers expect to return their catches. Trout anglers will need a greater area to fish in, whereas coarse anglers are happy to fish from one location. A greater number of coarse anglers can therefore be accommodated on a given length of bank and the overall economics usually favour coarse fisheries.

The fisheries manager must make his choice after considering all these factors.

2.1 Management of coarse fisheries

All fish are cold-blooded. Their body temperature is therefore more or less the same as the water temperature. As activity is related to water temperature, feeding and growth occurs mainly during the summer months in temperate zones (see Section 1.4). Fishing, for many species, is therefore often better at this time.

All the natural and semi-natural coarse fisheries contain some or all of the species mentioned in Section 1.4. The techniques used to manage these fisheries exploit their natural biological and physical requirements.

Types of coarse fish

In Section 1.4 coarse fish were divided into three groups. From a management point of view it is helpful to further subdivide the groups as follows:

Still water (or slow-flowing water) cyprinids

This group includes carp, tench, bream, roach and rudd. The adults live in slow-flowing rivers or lakes that are normally warm in the summer. The fish are predominantly shoal members, and spawn in spring on to early summer at temperatures varying from 12°C for roach to 20°C, or even 22°C for carp and tench, depending on the length of time over which the water warms up. The eggs are normally deposited on submerged plants in reed beds or on submerged tree roots such as alder. Although the fish spawn in shoals, the eggs are scattered fairly diffusely over the weeds so that some escape predation. The eggs hatch after about 3 weeks in the case of roach or 5 to 7 days in the case of carp (depending on temperatures).

The fry feed during hours of daylight on the early summer zooplankton. They grow quickly during the warm months and form large shoals at this time. The fish normally mature after 3 or 4 years. Some vegetable matter may be consumed along with bottom living invertebrates. In the case of carp there are a number of different scale patterns and body forms, king carp, leather carp, common carp and mirror carp are all the same species whereas crucian carp (without barbules at the mouth) are a different species.

Flowing water cyprinids

The second group of fish comprises the mid-river cyprinids such as barbel, dace, chub and bleak. These fish are naturally absent from stillwaters but can survive in reservoirs if introduced, although they rarely spawn successfully. They differ significantly from the carp and tench group in their temperature requirements. They spawn from March (in the case of dace) until June (for barbel and chub) at temperatures of 10–16°C. The preferred spawning substrate is gravelly shallows, and the eggs can take up to 25 days to hatch at 13°C. The fry are longer and thinner than carp or tench fry and can measure up to 7.5 mm. Maturity occurs after 2 or 3 years. Chub and barbel eat a great range of invertebrate matter as well as small fish and crayfish.

Predatory fish

The third group contains the predators pike, perch and zander. These are all carnivorous in habit, although the young stages take members of the zooplankton. Older pike and zander eat mainly fish, but perch eat a more varied diet of small fish, crustaceans, beetles and worms. Perch and zander form shoals to locate and catch their prey, but pike are solitary hunters. Pike spawn as early as February through to April in shallow areas on last year's reed beds: perch spawn after pike and leave very characteristic ribbons of eggs around reed stems; zander spawn from April to June at temperatures around 15°C, and the eggs are shed in shallow water on reed or rush stems.

Pike and zander grow large, pike reaching 20 kg and zander 8 kg. Perch are smaller fish, seldom exceeding 2 kg when mature.

Miscellaneous and small species

This fourth group consists of gudgeon, stoneloach, bullhead, minnows, and sticklebacks. Of these the gudgeon is quite important in fisheries such as the Trent and Soar rivers, and may also be found in canals and some stillwaters. Gudgeon, stoneloach and bullhead are all bottom-living and eat benthic invertebrates such as chironomid larvae and molluscs. Spawning takes place in April–May in dense shoals, and the sticky eggs are shed onto plants and stones. They take up to 2 weeks to hatch and growth rate is fast over 3 to 4 years, but slow thereafter. The biggest fish of this type are around 15 cm, but they are usually much smaller fish.

Creation of coarse fisheries

The following types of coarse fisheries can, to a great extent, be created according to the needs of the anglers fishing that particular pond fishery.

General purpose non-specialist coarse fishery

Most angling clubs and day ticket fishery owners will be looking for an all-round fishery that performs well throughout the year, i.e. contains fish that can be caught throughout the year and fish that can be caught by the inexperienced as well as the more experienced angler. The construction of an appropriate pond is dealt with in Section 2.3 and Appendix 5.

Ideally a period of 2 to 3 years (two or three growing seasons) should be allowed before a new lake is stocked, though financial pressures may shorten this. This period is needed for plants to become established and natural food to build up. If this period is foreshortened care should be taken to ensure that excessive numbers of fish or fish of large size are not introduced too soon or the fish may lose weight and condition. Anglers' baits will help supplement natural food and supplementary feeding can be undertaken if necessary during the warmer summer months. Unless floating pellets are used feeding should be done at a feeding station so that a check can be kept to prevent the accumulation of uneaten food. Normally food weighing about 1% of the weight of stocked fish should be fed daily if temperatures and oxygen levels are high.

Education, either formal of informal, is often needed. It may be necessary, for example, to have rules banning the use of keep nets in order to prevent unnecessary damage to fish. Disease problems are described in Section 1.7, and it should be borne in mind that while problems can often be prevented they can rarely be cured. An example of this is the outbreak of the viral disease Spring Viraemia of Carp (SVC) caused by the virus Rhabdovirus carpio, in the late 1980s. The virus, thought to be new to this country, was reported to be killing carp. SVC caused losses of carp at individual fisheries varying from 10% to 100%. Losses occurred in both cyprinid and ornamental fish stocks. The evidence suggested that imported fish were the most likely source of infection. SVC became a notifiable disease (MAFF must be notified of the presence of certain diseases in fish farms or in the wild. There are 11 such notifiable diseases, most of them appearing at fish farms only) and fisheries managers were offered advice by MAFF, including a disinfection technique. MAFF also produced a booklet on SVC obtainable free of charge from their London address (see Appendix 3 and their Publications Address, MAFF Publications, Lion House, Willowburn Estate, Alnwick, Northumberland, NE66 2PF).

Rod licences are required by all anglers, including the fishery owner, except juniors less than 11 years old. Unpleasant scenes can often be avoided if this requirement is printed on the permit, together with the other rules that apply at the fishery. This latter aspect is dealt with in Section 2.9.

Although anglers will suggest or even demand that a great variety of fish species be stocked, it is better in a smaller lake, e.g. of about 1 ha, to stock one chosen bottom-feeding species such as carp along with a midwater or surface species such as roach or rudd. Perch may be introduced to control the excessive reproduction of the roach and carp but in a non-specialist fishery predators such as pike or zander may get out of balance and are generally best avoided.

Stock can be introduced at most times of year but fish are likely to be badly stressed if transported on hot summer days and may suffer temperature shock if they are brought from water at a higher temperature and introduced into cooler waters, or vice versa. An acclimatization period is required.

There is generally little problem in fishing for introduced fish soon after they have been stocked and in waters where natural food is limited anglers' baits may be an important source of food.

Though carp may not spawn successfully every year if summer temperatures are low, for most coarse species a surplus of small fish is to be expected after a few years. These are best removed, e.g. by netting, to ensure continued growth and condition in the remaining fish. If this is not done a large number of small stunted fish are likely to remain in the fishery and further successful spawning will be inhibited.

An important part of the management of any fishery is the 'management' of the angler, in order that problems are avoided with the fish stock. For example, research in Holland in 1992 shows that if anglers use ground bait, catches in a mixed fishery (excluding carp) will not increase if an angler introduces more than 2 kg of bait in any one session. In small pond fisheries there is some justification for not allowing anglers to bait their swims with more than 2 kg of ground bait. Excess baiting can have a negative effect on oxygen conditions and the invertebrate fauna.

Specialist coarse fisheries

There appears to be an increasing demand by anglers for specialist coarse fisheries. Generally this will mean a fishery with good numbers of large fish of one particular species. Fisheries of this type need to be biologically productive and the chosen species stocked at a low initial density to allow for growth. To maintain a fishery of this type, regular cropping of any progeny from the initial stocking of fish must be undertaken in order to reduce competition for food and create conditions for growth.

The most common examples of specialist fisheries are those managed for carp and pike, but others include tench, zander, and wels (catfish) and eel. It should be noted that consent under the Wildlife and Countryside Act 1981 is required from MAFF, as well as NRA consent, before 'exotic' species such as zander, wels and grass carp can be introduced to UK waters. A fee is payable and application forms are available from MAFF, 17 Smith Square, London. At the present time there is no charge for consent under the Salmon and Freshwater Fisheries Act 1975, available from the appropriate NRA Region.

Carp fisheries

Carp (Fig. 2.1) were originally introduced into Britain around the twelfth century. Because their natural range is in areas of warmer summer climate, British water temperatures seldom reach the optimum for growth of this species (around 25°C). For this reason carp normally grow faster in shallow lakes than in deeper colder lakes. However, an ideal carp pool should have some deeper areas as well as shallow ones, since deeper water will buffer the effects of sudden changes in temperature.

Both large carp fisheries and small carp fisheries are encountered.

Large carp fisheries contain 'doubles' or 10+ 1b (4.5 kg) fish, 'twenties' or 20+ lb (9 kg) fish or even 'thirties', i.e. 30+ lb fish (13.6 kg). A pool that is to be developed as such a fishery may already contain a mixed fish population. If so, the first step is to remove the unwanted coarse fish, especially other bottom feeding species such as tench and bream. To do this properly, the pool should be drained either by lowering via a sluice or 'monk' (if present), or by pumping, (consent for draining ponds/lakes is required from the NRA and English Nature. It is also prudent to contact the local County Wildlife Trust).

Having obtained consent, it is wise to restock the pool with only about 150–250 kg/ha in order to allow for growth. If the pool was not very productive initially, fertilization should be undertaken to improve the productivity (see Section 2.6, page 85). If, after fertilization, the pool is capable of supporting

Fig. 2.1 A fine specimen-sized carp weighing 7.0 kg.

600 kg of fish per hectare, the young fish of successive spawnings will quickly grow to achieve the maximum biomass the pool can support. Annual or biannual cropping of small carp may then be necessary if good growth rates are to be maintained. An increase in weight of at least 1 kg per year is to be expected for carp over 0.5 kg. Under ideal conditions increases in weight of up to 5 kg per year can be recorded for well fed carp in productive understocked waters.

The cost of creating a specialist carp fishery by stocking large carp should not be underestimated. Fish, especially those over 9 kg, are difficult to buy and are likely to be very expensive. Because of their value the likelihood of their being stolen is also a significant factor and the security of the fish is a management consideration that should not be overlooked.

Small carp fisheries are ones that are deliberately overstocked with carp of a small size to create what the manager hopes will be an easy fishery for any angler and a productive match fishery. Because the carrying capacity of the lake is exceeded in terms of the natural food available the carp are very dependent on anglers' bait as a source of food. Adverts for such waters may include such phrases as 'thousands of hungry carp just waiting to be caught'. As a result of overstocking, problems are often encountered in a fishery of this sort and they are generally not to be recommended. The problems arise because the fish are overcrowded, undernourished, slow growing and stressed. Because of overcrowding and poor conditions parasites and diseases are often easily spread from fish to fish and the fish are less able to cope with these additional problems. The anglers and the fisheries manager are then open to criticism for not giving first priority to looking after the health of the fish in the fishery.

Tench and bream fisheries

The same general principles apply to those discussed above for carp but bream generally prefer somewhat deeper, more open water than carp. Tench are slow growing, particularly in their earlier years, and favour weeded areas (including blanket-weed) for spawning and fry survival.

Pike fisheries

The same basic techniques are required to produce a specimen pike fishery (Fig. 2.2). The pool should be fertilized (see Section 2.6) to provide a high standing crop of prey fish which in turn will support a high standing crop of pike.

Although carp fisheries can be developed in small waters, pools larger than about 3 ha are needed to produce this level of stock that will support a viable specialist pike fishery. The main problem with a pike fishery is achieving the

Fig. 2.2 A specimen-sized pike, weighing 10.0 kg.

correct balance between pike and prey fish. Pike often live in balance with a prey fish population when the weight-for-weight ratio is 1:7 or 1:8 – that is, 1 kg of pike for every 8 kg of prey fish, although there is evidence that balance can be achieved even at ratios as far apart as 1:4 and 1:20. Obviously the lower ratio is more advantageous to pike fishermen since any water stocked in this way will contain more pike. It should be borne in mind that a specialist pike fishery should, if managed correctly, also produce big fish of the prey species, as predation reduces their numbers, reduces competition and increases growth.

The type of prey introduced can be determined by the other types of fishing required and in larger water the prey fish stocks could comprise a mixture of bream, roach and/or rudd. A few pike fisheries have achieved satisfactory prey ratios by regularly stocking with rainbow trout.

Once the fishery has been created care must be exercised in managing it. Because the pike are likely to be caught several times (a tagged pike in a Midlands fishery was recaptured just under 30 times over two years) it may be necessary to persuade anglers to use tackle and tactics that cause the least damage to the fish they catch (barbless hooks, knotless nets, minimum line strengths). Similarly, in order to further protect stocks, the angler effort and use of the fishery may also have to be carefully controlled lest the accidental death of some of the fish leads to a deterioration in sport. When stocks are lost, additional replacement fish will be required. Stock fish can be obtained if contact is made with angling clubs who wish to remove unwanted pike and the necessary stocking consent obtained from the NRA.

Zander and wels fisheries

The same general principles apply as those described for the pike fishery above but it must be borne in mind that zander generally feed on much smaller prey than pike – typically roach of less than 120 mm. The prey species must therefore be carefully managed to ensure the regular production of large quantities of juvenile fish to support the predator stock. Habitat improvement, as described in Section 2.6, is therefore important, and particular attention should be paid to the improvement of spawning and fry habitat.

Eel fisheries

Eels are peculiar in the sense that, unlike the above species, they do not spawn in freshwater and will therefore need to be stocked in (or be able to migrate to) the fishery. Details of the stocking of eels in lakes are given on page 139. Most specialist eel fisheries are those from which the seaward migration is restricted or physically prevented. In this situation female eels in particular continue to grow to a large size, rather than undergo their normal spawning migration, which typically takes place in September or October. Continued light stockings of small eels (elver) will be required over the years to replace eels which are reaching the end of their natural life span. Despite stories of eels crawling over land, eels over 200 mm in length rarely if ever try to leave a lake other than in the outflowing water, which can be suitably screened to prevent this escape.

2.2. Management of salmonid fisheries

Six species of salmonids form breeding populations in the British Isles. Four of these – Atlantic salmon, brown trout, char and grayling – are native to this country; the other two – rainbow trout and American brook trout – were introduced in the last century and have established self-sustaining populations at a few locations. Of these six species the grayling is unusual in that it spawns during the spring as opposed to the typical salmonid spawning period of the autumn. Because of this the grayling is often classed as a coarse fish for fishing seasons but its adipose fin clearly identifies it biologically as a salmonid.

Atlantic salmon are migratory fish, as are some stocks of brown trout (sea trout), and their life cycle involves a riverine spawning and juvenile phase together with a marine growing period of varying length. Such a complex life cycle has certain implications for management of these species in that they are susceptible to different pressures at different times and in different locations. A further complication in the management of salmonid fisheries is that in most cases, fish caught by anglers or commercial fishermen are removed from the water for consumption or sale, a practice which can profoundly affect the fish population as the fishermen are acting as an extra predator upon the fish stocks.

Fishery managers can gain some idea of the suitability of a water for salmonids from the biological requirements presented in Table 2.1. There are specific management techniques for salmonid fisheries in still and running water and they may be conveniently divided into direct and indirect methods. Indirect methods tend to be more applicable to self-sustaining fisheries and are aimed at increasing or maintaining natural production of fish, whilst direct methods are more often encountered in maintained fisheries and involve straightforward manipulation of fish stocks.

Indirect methods

Increased spawning escapement

The number of adult fish able to spawn is obviously of fundamental importance to the viability of a fish population and some of the indirect management methods are intended to protect this part of the life cycle, especially as anglers and commercial fishermen exploit the adult fish which are the broodstock.

Table 2.1 Summary of biological and environmental needs of the main British salmonids

Requirement	Salmon-river stage	Brown trout (BT)	Rainbow trout	American brook trout	Grayling
Preferred growth temperature °C	13–15	12	14	12–14	10–16
Normal maximum temperature	16–17	19	20–21	19	20
Spawning temperature °C	0–8	2–10	4–10	2–10	10
Normal spawning times	November–December	November–December	November	Sept–November	April–May
Size of spawning gravel	3–16 cm	1–5 cm	1–5 cm	1–5 cm	1–5 cm
Egg size	5–7 mm	3–6 mm	approx. 5 mm	3–6 mm	3–4 mm
No. of eggs per kg of adult body weight	1100–2000	1100–2600	1100–2600	approx. 2800	6–10 000
Incubation period	88 days at 5°C	40 days at 10°C	32 at 10°C	85 days at 7°C	25 days at 18°C
pH range	5–9	5–9	5–9	4.5–9.5	5.5–8.5
DO requirements	> 5 mg/l best at 9 mg/l	> 5 mg/l best at 9 mg/l	4–4.5 mg/l best at 9 mg/l	4–4.5 mg/l best at 9 mg/l	> 5.0 mg/l best at 9 mg/l
Feeding requirements	Similar to BT but feed more often on surface and live in faster water	Feed on bottom animals, in mid-water and on surface	Similar to BT – thought to feed over a wider temp. range	Feeds in mid-water and from surface and at lower temp. than BT	Feeds at all levels and over a wide temp. range

Regulation of fishing effort by means of close seasons and close times, together with the prohibition of some fishing methods and locations, allows the escapement of spawning fish which may otherwise have been caught. In the case of angling, some size limit regulations, particularly 'slot limits' which permit the retention of middle-sized fish but demand the return of large fish, are designed to ensure adequate numbers of spawning fish. Normal minimum size limits are intended to protect the potential spawners of the future but may not prevent overexploitation of the current breeding stock.

Kelt conservation
Conservation of salmon kelts (fish that have spawned), by means of prohibiting their capture is a frequently used method of salmon protection. However, as post-spawning mortality is often as high as 95% and their contribution as repeat spawners is usually not significant, kelt conservation is of relatively limited value for most rivers. Reconditioning and release of wild broodstock from hatchery facilities may be more beneficial as the fish are normally in better condition than wild fish and probability of survival is higher.

Improved water quality
Salmonids are amongst the more sensitive fish species with regard to water quality and it follows that measures taken to control pollution of all sorts can be an important aspect determining fisheries quality. Catchment-wide pollution control is the responsibility of large agencies such as the NRA but on a smaller scale good management of farm waste disposal, pesticide and fertilizer usage can be beneficial for individual fisheries.

Increased availability of spawning areas

On river systems that have been used extensively as sources of power in the past, it is frequently the case that numerous weirs, mills and dams remain intact long after their functional life is over, presenting often insurmountable obstacles to migrating adult salmonids and preventing their access to otherwise suitable spawning areas. Most often Atlantic salmon and sea trout are affected in this way but non-migratory brown trout may also have their spawning migration impeded. As much as 45% of the potential spawning habitat may be inaccessible to spawning salmonids in some catchments.

There is a legal requirement to provide fish passage facilities in any newly-constructed weir or dam on rivers frequented by migratory fish but on existing obstructions there is no such obligation. Provision of fish passes (Fig. 2.3, a and b) on existing obstructions may be undertaken by the NRA as long as the benefit in terms of increased access to spawning facilities is seen to be significant. Most of these works are relatively complex and very expensive engineering projects and are beyond the means of most private fishery managers.

In some cases fish have access to sections of river that do not have the required type or quantity of gravels to permit spawning. When this occurs it may be possible to build spawning areas to suit different species of fish by importing gravel to a section of river or by building special off-stream

Fig. 2.3a Denil type fish pass.

Fig. 2.3b Pool and overflow type fish pass.

spawning channels. As with fish passes this sort of construction work is very expensive but has been shown to be effective.

Other improvements to existing spawning areas are possible and may include some of those techniques discussed elsewhere in this book and include the prevention or removal of siltation in the spawning gravels and the possible removal of predators from the spawning and nursery areas.

Land use in the catchment areas of spawning and nursery streams can have great influence upon their suitability. Probably the most influential land use in this respect is commercial forestry development, which has been shown to be a significant contributory factor in the problems of surface water acidification, a major deleterious effect upon salmonid habitat quality. Figure 2.4 shows a tributary being limed to counteract this. Afforestation and its associated land preparation methods also exert influence upon the drainage characteristics of the catchment and can result in excess siltation of spawning gravels and increased 'flashiness' of flows.

Direct methods

Direct methods of salmonid fishery management involve the collection and care of broodstock, the production and incubation of eggs and the rearing of fish to a suitable size. These methods have been used in many river systems in an attempt to improve Atlantic salmon and sea trout fisheries, with stocking out of fertilized eggs or young fish as fry, parr or smolts. The older and larger the fish at the time of stocking, the greater the proportion of adults subsequently returning to the river.

If stocking of salmon is to be undertaken then care must be exercised in the

Fig. 2.4 Dosing an upland tributary with powdered limestone.

selection of broodstock to protect the genetic integrity of the catchment or river system. Efforts should be made to obtain broodstock from the river to be stocked or from an adjacent catchment. Importation of 'foreign' broodstock to a river with existing salmon may well weaken the resident gene pool and result in the fish being less well adapted to that particular river.

The numbers of juvenile salmon or sea-trout that are stocked is crucial to the success of the operation. In assessing the stocking requirement it should be remembered that the number of fish that will live in a stretch of river (the carrying capacity) is governed by the physical characteristics of the river channel. Stocking beyond the carrying capacity will represent wasted effort as the 'excess' fish will die or migrate out of the system. Understocking is also to be avoided as it means that the stream is failing to reach its potential in terms of smolt production.

Stocking is a viable proposition where natural production is below capacity. For example, when a reservoir or power station is built and significant lengths of spawning stream become inaccessible or where water quality in a catchment has improved to the point where salmon runs may be re-established. Suggested numbers of the different life stages to be stocked in rivers with very little or no natural production are given in Table 2.2.

Maintained fisheries

For non-migratory salmonids direct management techniques are used in rivers

Table 2.2 Recommended stocking rates for salmon eggs and fry

Green ova, eyed ova, unfed fry	200–600/100 m² (in batches of 1–2000)
Fed fry	100–300/100 m²
1 + parr	40–120/100 m²
2 + parr	5–15/100 m²

and stillwaters to support and maintain fisheries which are subjected to higher fishing pressure than the natural productivity can withstand or to provide sport with much larger fish than the water would produce. Maintained fisheries may also be created in waters which have no natural salmonid population but are otherwise suitable.

Flowing water fisheries

In rivers the usual choice of species is brown trout. In streams and rivers with a wild population of browns it has been shown to be inadvisable to introduce rainbow trout as the resident fish often decline as a consequence. Brook trout may be suitable for river stocking in some circumstances. As a general guideline, it is usually best to stock with takeable-sized fish. The introduction of eggs, fry or under-sized fish to a water with a low number of anglers and poor spawning facilities is possible but will not produce consistent, high quality sport that is the accepted norm on many modern trout fisheries.

Stock fish are an expensive investment and careless handling of the stocking arrangements can lead to poor sport and financial loss. Experience has shown that in a typical trout river a natural production of between 5 and 20 takeable trout per 100 m of stream may be expected, depending on habitat quality and stream width. If the fishing pressure is expected to be such that the annual catch will exceed this figure then stocking will be necessary.

Stocking should be restricted to shortly before and during the fishing season as it has been shown that there is a much higher percentage return to anglers from fish stocked at these times than when fish have been stocked in the autumn preceding the fishing season. In rivers, overall returns can be expected in the region of 40–70%, depending on fishing pressure. Frequent stockings with small numbers of fish will also improve return rates and give more consistent sport because in many waters stock fish are caught rapidly, with up to 75% of the total catch being taken within 2 weeks of stocking. Wherever possible, introduction of new fish should be carried out at a rate commensurate with the current performance of the fishery but the need to make arrangements with the supplier over numbers and dates of delivery may preclude this fine tuning. Good quality angler catch returns are essential to the sound management of any maintained fishery as the numbers of fish caught or the catch per angler will determine when and how many fish need to be stocked.

Towards the end of the fishing season in rivers, the fisheries manager should carefully monitor catches and stocking to ensure that as few fish as possible remain uncaught at the end of the season. There is little direct evidence that 'leftover' stock fish contribute to the spawning activity in streams and over-wintering survival is likely to be relatively poor. Alteration or relaxation of fishing regulations (allowing bait fishing, increasing bag limits) in the last few weeks of the season may be beneficial in reducing the number of uncaught stock fish.

Stillwater fisheries

In stillwaters, salmonids very rarely reproduce successfully and consequently practically all fisheries in pools and lakes are maintained by regular stocking. All species of cultivated salmonid can be used in stillwaters but rainbow and brown trout predominate. Brook trout and Atlantic salmon have been used in some stillwaters with varying degrees of success but the main attraction seems to be one of novelty.

Rainbow trout are the most widely used stock fish as they are cheaper to buy, more tolerant of marginal water quality and reputedly easier to catch than brown trout, a characteristic that is supported by the generally much higher return rates of rainbows (85–95%, compared to 55–75% for browns). Over-wintering survival of rainbow trout is lower than that for browns, however, and this may be a consideration in some circumstances.

As with river fisheries, frequent stocking with small numbers of takeable-sized fish just before and during the season produces the most successful and consistent fishing. Appropriate stocking densities will vary from water to water, according to the type of fishing required, the expected fishing pressure and the size of the water. In very large reservoirs stocking densities in the range of 50 to 80 fish/ha have been found to be satisfactory but in smaller pools much higher densities are often used. Where fishing pressure is relatively low and stock fish are expected to live for some considerable time before being caught, it is wise to limit the stocking rate to the natural carrying capacity of the water. For most lakes and pools this will be in the region of 75–100 kg/ha, a biomass that can be supported by natural food production. This level of stocking will not withstand high fishing pressure, though, and most commercially viable fisheries operate on a purely put-and-take basis. Again overwintering survival of stock should be assumed to be low, when calculating the stocking rate for the following season.

In small waters stock fish are rarely resident for long enough to grow significantly and if very large fish are required by the anglers then they must be introduced. A small number of very large fish in a pool often stimulates the interest of anglers, and as a high proportion are caught, the investment is justified. Large stillwaters often produce very good growth rates of stock fish and are capable of generating their own 'specimens'.

Catch-and-release

One alternative to managing stillwater trout fisheries on a put-and-take basis is to operate a catch-and-release system, where captured fish may be returned alive to the fishery. Contrary to popular belief the mortality rate of released fish is relatively low if the fish are handled correctly, and individual fish may be recaptured as many as four times. Catch-and-release therefore provides very high return rates on stock fish and between 200% and 300% may be expected, effectively reducing the cost per fish caught by a significant degree. To ensure the best chance of survival of released fish it is advisable to fish fly only, to use barbless hooks and to reduce the handling of the fish to a minimum by unhooking whilst still in the water. Rainbow trout in this type of fishery tend to be caught for up to 3 years before dying of 'old age'; browns are much longer lived and may appear in catches for 7 or more years after stocking.

2.3 Construction of stillwaters

There will be occasions when the only way to obtain a fishery is either to create an entirely new one, or to restore an existing pool. Anyone undertaking this work will need to take into account the many legal requirements which could apply (see Appendix 7).

Requirements for coarse fish pools

If it is intended that the pool be a coarse fishery, great water depths are not required; a good design would be one where one-third of the pond was less than 1 m deep, one-third of medium depth (1–2 m deep) and one-third 2–3 m deep (see Fig. 2.5). Coarse fish generally favour warmer water than trout, and shallow waters are the most productive for food. Although there are advantages in having very shallow margins, they may not be fishable and it is better if these are dug out to 1–2 m deep. This helps to prevent excessive growths of marginal vegetation and enables keep nets to be used. It also provides more water area for fishing. Where possible, the pool should have indented margins which will add more variety to the pool so that anglers may find sheltered spots, and will increase the bankside space available. Screening trees can also be planted (see Section 2.7). There are different views on the value of constructing islands. Islands can cause difficulties in small pools but can be a feature in larger pools and provide a refuge for wildlife. Submerged islands are best in some coarse fisheries. They can provide a shallow area for aquatic plants to grow away from anglers, allowing them to fish out to the shallows.

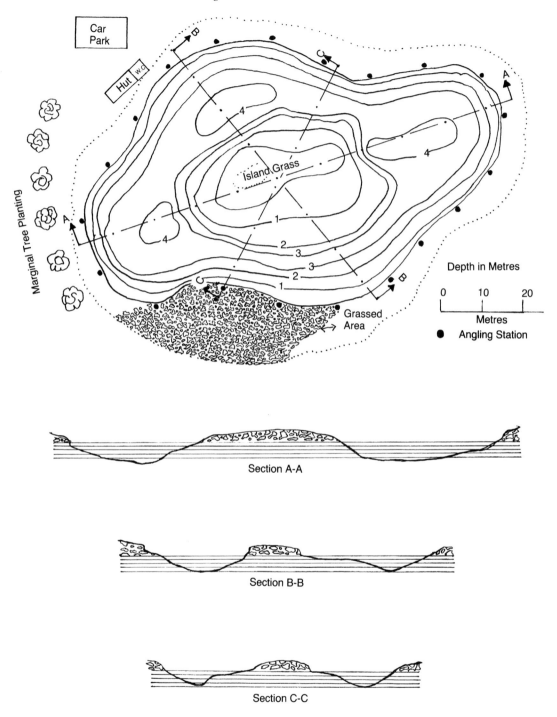

Section A-A

Section B-B

Section C-C

(After R. Millichamp. IFM Pond Booklet 1989.)

Fig. 2.5 Typical coarse fishery, plan and profile.

Requirements for trout pools

In general terms, trout pools should be deeper than those described for coarse fishery. Trout do not thrive in temperatures in excess of 20°C, and in a hot summer this temperature may be exceeded in a shallow pool. An average depth of 2.5 m is ideal, with some areas exceeding 3.5 m. Shallow areas should be greater than 1 m, and the margins should be relatively deep (1–2 m), although this is not essential. A variation in marginal depth gives suitable habitat for marginal plants and allows anglers to fish without casting problems. Lakes of less than 0.5 ha are not normally suitable for trout unless there is a flow-through of fresh water.

Excavation

Excavation of a pond or lake is an expensive business so care should be exercised to ensure that the most economical equipment is used for the job (see Appendix 2). It is essential that the site is surveyed, and the depths determined, so that deep areas can be constructed where there is little soil to remove, and 'high spots' are left as shallows. Part of assessing whether the site is suitable (sometimes overlooked) is to ensure that access to the site is possible for hydraulic excavators, etc. It is a good idea to dig a few trial holes across the proposed bed of the lake to determine the nature of the bed before work begins.

If the pool is to be constructed in the washlands or floodplain of a river then consent of the NRA will be required. No raising of ground levels will be permitted so the spoil will need to be moved off site or mounded out of the washlands. This can add appreciably to the costs of digging a pond.

Care should also be taken when choosing the site to avoid digging the pond in important wetland sites or unimproved meadow land, which may have a high conservation value.

If the pool is to be constructed by damming a valley, then great care should be taken. A pool of average depth 2.5 m and area 1.2 ha impounds about 25 000 m^3 of water, and at this size and larger, the construction of the reservoir requires a 'Panel 1' civil engineer to supervise the design and construction. (Such reservoirs must be registered with the local authority.) This can prove very costly. The economics indicate that unless a lake of nearly 4 ha is required, it is better to confine the area to 1.2 ha and/or the capacity to below 25 000^3. If in doubt, seek the advice of a qualified engineer. (A 'Panel 1' civil engineer is one who, under The Reservoir Safety Act 1975, has been appointed by the Secretary of State to Panel 1 of the three Panels of Reservoir Inspectors.)

Construction of on-line pools

There are several major points to remember when constructing a pool, some of which are shown in Fig. 2.6.

If a pool is to be constructed on-line (that is with a stream flowing into it and overflowing out) it will be liable to silting and may require dredging after a number of years.

The overflow or spillway of the pool should be wide enough to allow flood flows to pass over it to a depth of no more than 7.5 cm. This will prevent loss of fish and will eliminate the need for gratings, which require regular cleaning. (Details of flood flows can be obtained from a flood defence engineer of the NRA.)

Calculation of spillway size

Method 1: for small ponds
If the highest river-bank marks left by a flood were 3 m apart, measured from bank to bank, and the average depth of water during the flood was 0.5 m, this is divided by 10 cm (desired depth of flow) to give a factor of 5. The 3 m width is then multiplied by 5 to give a spillway width of 15 m which will be sufficiently wide to handle water from a flood of the size estimated. This method works satisfactorily only if the drift deposits can be located along the stream banks, their height measured, and there is no obstruction downstream from the pond.

Method 2: for drainage areas of less than 20 ha supplying a pond
The total number of hectares in the drainage catchment is divided by two to give an arbitrary spillway width in metres. To this is added a further 3 m as a safety margin. If, for example, the drainage area of the pond is 11.2 ha, this method will determine that the spillway should be 8.6 m wide.

Method 3: for drainage areas of more than 50 ha
The size of the spillway may be computed from the amount of 'run-off' from the drainage area. Specialist advice will be necessary to calculate this figure.

Location of spillway

Since most ponds are built in natural hollows, the spillway is often constructed on the dam. However, it may be located at one or both ends of the dam, or at a convenient point alongside the pond.

Spillway construction

The spillway should be paved with stone or concrete to prevent erosion. The

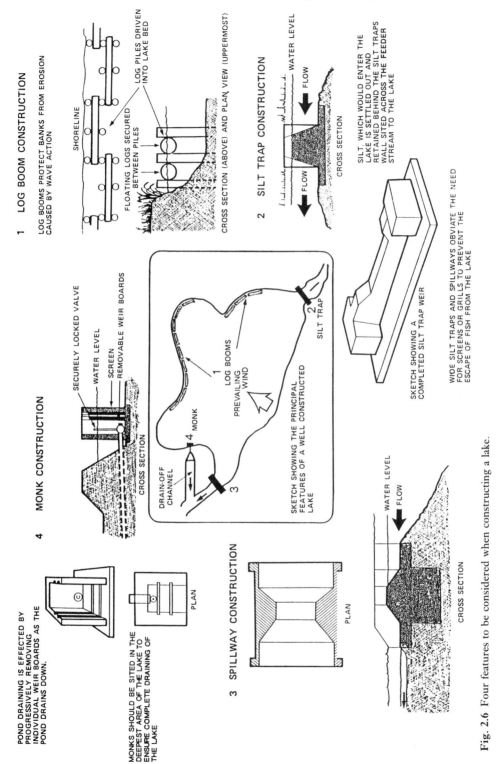

1 LOG BOOM CONSTRUCTION

LOG BOOMS PROTECT BANKS FROM EROSION
CAUSED BY WAVE ACTION

SHORELINE

LOG PILES DRIVEN
INTO LAKE BED

FLOATING LOGS SECURED
BETWEEN PILES

CROSS SECTION (ABOVE) AND PLAN VIEW (UPPERMOST)

2 SILT TRAP CONSTRUCTION

WATER LEVEL

FLOW

FLOW

CROSS SECTION

SILT, WHICH WOULD ENTER THE
LAKE IS SETTLED OUT AND
RETAINED BEHIND THE SILT TRAPS
WALL SITED ACROSS THE FEEDER
STREAM TO THE LAKE

SKETCH SHOWING A
COMPLETED SILT TRAP WEIR

WIDE SILT TRAPS AND SPILLWAYS OBVIATE THE NEED
FOR SCREENS OR GRILLS TO PREVENT THE
ESCAPE OF FISH FROM THE LAKE

4 MONK CONSTRUCTION

SECURELY LOCKED VALVE

WATER LEVEL

SCREEN

REMOVABLE WEIR BOARDS

CROSS SECTION

POND DRAINING IS EFFECTED BY
PROGRESSIVELY REMOVING
INDIVIDUAL WEIR BOARDS AS THE
POND DRAINS DOWN.

PLAN

MONKS SHOULD BE SITED IN THE
DEEPEST AREA OF THE LAKE TO
ENSURE COMPLETE DRAINING OF
THE LAKE

DRAIN-OFF
CHANNEL

4 MONK

3

1

LOG BOOMS

PREVAILING
WIND

2
SILT TRAP

SKETCH SHOWING THE PRINCIPAL
FEATURES OF A WELL CONSTRUCTED
LAKE

3 SPILLWAY CONSTRUCTION

PLAN

WATER LEVEL

FLOW

CROSS SECTION

Fig. 2.6 Four features to be considered when constructing a lake.

type of construction to use will depend on the location of the spillway, the type of soil on which it is to be built, and the amount of water it will have to carry. Regardless of the type of construction used, the spillway that is to cope with all overflow should be paved for a sufficient distance to carry the water away from the dam to avoid any erosion to it.

Diversion ditches

An on-line pool may be protected from flood flows by constructing a diversion ditch around the pool. Once the design has been made and construction completed, the ditch should be seeded so that its grassed bed will reduce the eroding effect of the flood water. A diversion dam or sluice can then be constructed in the stream above the pool so that flood flows are forced around the pool and down the ditch. One suitable design might be an earth dam built across the stream and incorporating a pipe which would discharge into the pool and govern its supply. Alternatively, sluices could be sited in the stream in order to allow a supply to the pool, but enable this to be shut down in times of flood.

Erosion

On large ponds, the wind often creates waves large enough to erode windward banks above and below water level. This action on new ponds may be so severe that a quarter of a metre or more of bank can be washed away in one day. A single log wall built along the bank at normal water level will absorb the impact of the wave's energy and prevent this erosion, as will a gently sloping bank lined with reeds. Logs 25–30 cm in diameter and 6 m long are placed in position along the face of the bank (see Fig. 2.6) and stakes are driven into the bed at each side to hold them in place. The logs should be lapped so that there are no gaps between them. A system of this type allows a certain amount of fluctuation of water level whilst still maintaining good bank protection. Floating log booms will reduce wave build-up across large stretches of open water.

On small ponds, a short-term solution is to place a good layer of turf on the face of the bank or dam to help prevent serious erosion by wave action.

Another method of reducing erosion is to drive willow stakes into the bank below water level and weave willow branches in and out of the stakes. The gap behind the stakes is then infilled with soil and either seeded or covered with turf. This method has the advantage that the willow stakes and branches will often grow, providing shelter and cover for fish.

Off-line pools

Pools constructed off-line do not usually suffer from the same silt problems as those on-line. The water supply to the pool can be obtained from a diversion channel, and the inflow may be controlled by a sluice or penstock at the junction of the brook, or by a small dam containing a pipe. Flood flows can safely pass down the brookcourse. A pool of this type will not require a large spillway, since flood flows will pass around it.

Silt trap construction

If the pool is fed by a stream, it is often desirable to build a wall across the stream to isolate an area near the inlet so that this will act as a silt trap (Fig. 2.6). Alternatively, a small pool can be constructed in the stream upstream of the fishing pool, and this can be cleaned regularly to reduce siltation in the fishery itself. To be as effective as possible, the silt trap should have a long weir apron over which the water passes, as this slows the water velocity and prevents silt from being carried downstream.

Draining pools

If a stillwater fishery is to be managed correctly, it may be necessary periodically to remove unwanted fish stocks, and since pools decline in productivity as they age, it is also beneficial to leave them fallow occasionally.

There are several means of assisting pool drainage, and the simplest type, if the bed of the pond is higher in level than a convenient nearby stream, is a monk, named after monks who first devised the structures for their fish ponds (see Figs. 2.6 and 2.7). A monk consists of a box containing a pipe at the bottom

Fig. 2.7 A concrete monk.

which drains the ponds. The depth of water is governed by stop-boards, which can be raised or lowered at will to control the height of the water in the pond. It is best to site the monk away from the bank so that it may only be reached by boat, as this will help prevent unauthorized tampering. If the monk is constructed of wood, timber such as pitch pine or elm should be used since they can withstand about 30 years immersion in water without rotting. Concrete collars can be used with concrete slabs as stop-logs if a more permanent structure is to be built.

Legal requirements

The erection, alteration or repair of any structure in, over or under a watercourse designated 'main river' will require the consent of the NRA under the Water Resources Act 1991.

Development of any kind adjacent to a 'main river' may require consent under NRA land drainage byelaws. Any obstruction (weir, mill, dam, etc) to flow or any culvert to an ordinary watercourse will require the consent of NRA under the Land Drainage Act 1991.

It is wise to seek the advice of the NRA on any development adjacent to any watercourse. Consent may also be needed from the Planning Authority. Informal discussions are often the best way of finding out whether legal consent is required. What may require formal approval in one area may be permitted without it in an adjoining area (see Appendix 2).

The design and construction of water-retaining structures is a skilled job requiring professional expertise. The failure of an impounding structure, even though it may retain only relatively small quantities of water, can have severe consequences and may put lives at risk. It is therefore recommended that the design and supervision of construction of impounding works be carried out by a qualified and experienced civil engineer. It is worth remembering that land suitable for creating a lake may not be under the same ownership as the watercourse itself.

2.4 Maintenance of stillwaters

Over a period of years there is a tendency for ponds to revert to marshland, but this process can be arrested if the silt that accumulates within them is removed. This is particularly true when there are large deposits of black, semi-decomposed leaf-litter and debris. This material will tend to reduce oxygen levels in the pond, and because it often produces acidic conditions it may also reduce the fertility of the pool. In pools that have become very shallow as a result of silt and organic debris, fish stocks may be at risk, mortalities may occur, and the maximum standing crop that can be sustained may be reduced.

Desilting techniques

Desilting a pool can prove costly, but costs can be significantly reduced if spoil can be deposited 'on site' rather than transported elsewhere. The silt may not be mounded adjacent to the pond if it is in the washlands of a river. Care should be taken to avoid depositing silt around trees as it can suffocate the root system. Wetland areas should not be used for silt as these can be important sites for wildlife. The silt removal exercise will be easier if the pond can be drained; on a pool such as a gravel pit without any monk or sluice arrangement, it may be possible to pump out the pool. This will depend on the permeability of the adjacent land and the proximity of water sources, but it enables the operator to see what he is doing, and the spoil is relatively dry and more easily handled. Care should be taken, however, to ensure that muddy water does not pollute watercourses downstream.

Types of machine and their limitations

The types of machine that can be used to desilt a pool (Figs. 2.8 and 2.9) will vary from pool to pool according to local topography, strength of substrate, ease of access, bankside vegetation and a number of other factors – including cost.

Draglines

The dragline is the most common type of machine used for larger stillwaters, but it has limitations. It is rarely possible to use it from anywhere other than the bank, and the maximum reach that can be achieved is normally 20 m, although a very large machine may be capable of reaching over 30 m.

For safety reasons, draglines cannot operate within $1\frac{1}{4}$ jib lengths of an overhead power line, and they require a considerable clear area in which to swing the bucket and dump spoil. They cannot therefore be used on a pool surrounded by trees, unless some of these are removed. Draglines are usually delivered to the site on a large trailer which requires good access; they require a track width of at least 3.0 m and because of their extreme weight, small bridges may not be able to accommodate them.

Hydraulic excavator

The reach of a hydraulic excavator is limited and this type of machine is not generally suitable for the majority of pools. On small pools that are surrounded by trees, however, this type of excavator may be the only practical means of removing silt. Larger machines can have jib lengths of over 15 m reach.

Drott

A drott can only be used when access to the pond bottom is possible, and

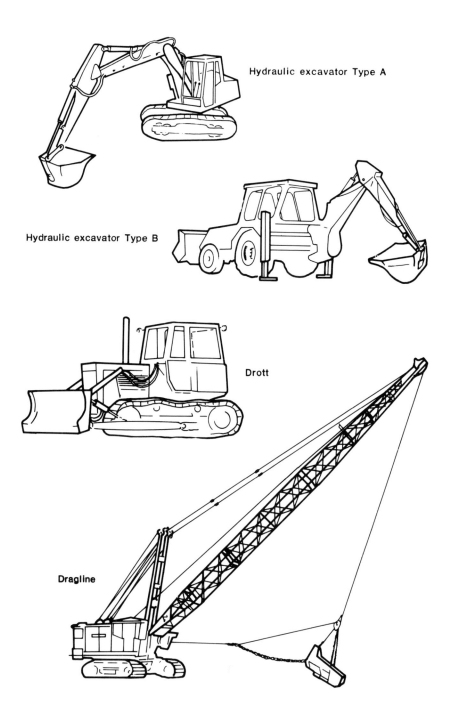

Fig. 2.8 Types of excavators used for stillwater construction and dredging operations.

Fig. 2.9 Dredging operation in progress.

where there is a solid base to the pond under the silt. Drotts can move the spoil and lift and dump it. They are generally used for landscaping work around pools.

Bulldozer

A bulldozer can only push soil, and requires access to the pond bottom which should therefore be firm under the silt. This type of machine can quickly move a large amount of spoil, and can be fitted with swamp tracks to prevent it sinking into silt.

Small punt-mounted excavator

This has a minimal reach and it is difficult to dispose of the spoil efficiently on the banks since the boat has to make repeated trips to the shore. With this machine it is possible to work nearer to power lines than with a dragline, and it can often reach parts of the pool that other excavators cannot (Fig. 2.9).

Mud cat

The mud cat works by sucking up silt from the lake bed through flexible pipes, and discharging it to specially constructed reservoirs on land. Its main disadvantage is that it is extremely costly to bring to site and, because of its size, it requires good access. It pumps away a great deal of water as well as silt, and there must be sufficient space around the pool for the construction of settlement lagoons (which often take more than a year to dry out thoroughly and consolidate).

It is usually best to obtain a quotation for the total cost of the job rather than hire a machine on an hourly basis; this is because the driver in these circumstances will require supervision and instruction while working and this may not be practicable for most fishery managers.

2.5 Control of aquatic plants

Water-plants are found in almost all aquatic situations in Britain, and are an essential part of the biology of rivers and lakes. A simplified summary of the role of plants in a fishery is shown in Fig. 2.10. Aquatic vegetation is important because it:

- Aerates water by the process of photosynthesis. Much of the oxygen produced is removed by respiration during the day by animals, and at night by both plants and animals. During very warm weather the respiratory demand for oxygen may be so great that it falls to a level critically low for fish.
- Acts as a shelter for animals. Larger plants provide a refuge for animals, protect them from fast water currents, and hide them from predators. In addition, many species of fish and invertebrates deposit their eggs on

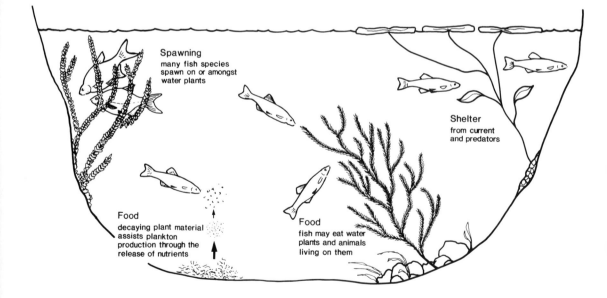

Fig. 2.10 A summary of the role of water-plants in a fishery.

plants. The presence of aquatic vegetation increases the productivity of a fishery, and a water that is rich in plant-life will generally contain more food for fish than one that possesses an impoverished flora. Too much vegetation can, however, lead to water quality problems and make angling difficult.

- Consolidates the beds and banks of a fishery. Plant roots help to bind the soil and prevent the banks from collapsing; rooted plants anchor gravel and stones, so making the bed more habitable for invertebrates, and vegetation absorbs the impact of current and waves and protects banks from erosion.
- Provides food for other organisms. Plants, whether living or dead, form a food source for many animals. Plant detritus, for example, is a major food item for many invertebrates, and aquatic vegetation is an important part of the diet of many cyprinid fish.
- Intercepts silt and plant debris. Plants reduce water velocity and raise water levels when they grow in running water. They may thus assist the deposition of silt and organic matter in some parts of the channel, and increase water velocity in other parts, thereby creating scour. In stillwaters, decomposing plant material can accumulate over the years. This increases the productivity of the pool and the density of growing plants, and helps to raise the bed, thus promoting the development of marshes and (ultimately) dry land.

Types of aquatic vegetation

Aquatic plants (see Appendix 6) can be classified into four groups:

Emergent and marginal plants
These plants possess erect, narrow leaves and will be rooted mainly in standing water or on the adjacent bank. Tall grasses such as reed sweet grass can dominate marginal vegetation. Reeds, grasses, sedges, rush and reed mace are all found in this group.

Floating leaved plants
This group includes water lilies. Most of the plants are rooted in the bed and have long pliable stems, but a few, such as duckweed and frog-bit, drift on the water surface. Members of this group are often found growing together with emergent and submerged plants in water just over 1 m deep. The free-floating species are found over virtually any water depth.

Submerged plants
These are commonly rooted in the mud, like Canadian pondweed and water milfoil, but a few are free-floating below the water surface, like ivy-leaved duckweed and hornwort. They are all completely submerged except at the

flowering period when most extend their flowering shoots above the water (e.g. water milfoil and mare's tail).

Algae

Algae are primitive plants which are classified botanically according to the colour of pigment they contain. Filamentous green algae are common examples, often growing in entangled mats and known as 'blanket weed' or 'cott'. The group also includes many microscopic forms that float about in stillwater and give rise to dense 'blooms' when conditions are suitable for their rapid growth and multiplication.

Unicellular blue-green algae can be present in any water. They tend to bloom when there are insufficient macrophytes in the waterbody to use up nutrients. The algae can be toxic to fish, man and animals and it may be necessary to close a water used for recreational purposes if certain types of blue-green algae are present.

Blue-green algae can be controlled to some extent with the herbicides terbutryne and diquat. Silver carp eat the algae to some extent. A considerable amount of research is taking place on the use of well rubbed barley straw introduced into the water in spring. An algistat in the straw prevents the algae developing.

Establishment and growth

Most troublesome plants are perennials: they die back in the winter and reproduce vegetatively each year from stems or rhizomes (underground runners). Some water milfoil produce turions, compact winter buds that drop to the bottom in the autumn and re-grow in the spring rather like seeds. Algae reproduce vegetatively from their own strands, or by producing spores.

Many of the perennial water plants re-grow very rapidly once growth has started in the spring. Water crowfoot, for example, begins growth in April or May and stems may reach lengths of 6 m by the end of June. Re-growth after cutting in the summer is equally rapid if the plant is cut before it flowers, but is slower if cut after flowering.

Objective of control

It is clear from the foregoing that aquatic plants not only play an important role in the environment, but that they also have a variety of forms. Fisheries management should not aim to eradicate aquatic vegetation, but to control it, and this can be achieved in a number of ways.

Cutting

Cutting is one of the commonest methods of weed control, and operations usually start in May. In shallow rivers, plants can be trimmed by using hand scythes, but where the depth of the water makes this impossible, chain scythes may be used. Cutting at the wrong time of year can stimulate fast re-growth. Each plant has a different growth pattern, and more efficient and economic control can be achieved if cutting times are based on these growth patterns. Two operators are necessary to use a chain scythe; each man operates from his own bank, and the scythe is moved back and forth whilst moving upstream. A similar method can also be used on lakes. When cutting is carried out on a river, a stop-net should be used to prevent plants floating downstream and causing a nuisance to other water users. Cut plants must then be removed from the river. The weed cutting methods for a chalk stream, described by Richard Seymour, are shown in Fig. 2.11.

For use on larger waters, a small, U-shaped mechanical cutter (Fig. 2.12) is available for attachment to a dinghy. Introduced to Britain from the USA, it cuts a swathe about 1 m wide to a depth of about 1 m. The cutting blade is operated by a small engine mounted on the top of the knife frame. A rake attachment, which enables the boat to be used to collect the cut weed in standing water, is also available. It is also possible to purchase larger purpose-built weed-cutting boats, but these are expensive. Other means of cutting water weeds include excavator-mounted weed-cutting buckets for use on small watercourses.

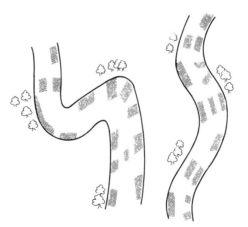

Copyright by Charles Knight & Co. Ltd. (1970). From R. Seymour.

'Fishery Management and Keepering'.
Cutting river weed by the Side and Bar method.
– Proper cutting on bends can prevent lateral erosion of banks

Fig. 2.11 Weed cutting in a chalk stream can control water flow and reduce bank erosion.

Fig. 2.12 Specially designed weed-cutting boats can be used to control weed growth in larger watercourses.

Raking or 'cotting'

Rakes are used to remove 'cott' from the water surface, and floating rope booms may be used to draw floating plants such as duckweed to the sides of a water where it can be removed. Rakes can also be employed to remove submerged aquatic plants like Canadian pondweed. Raking is a slow and labour-intensive technique, and is only really suited to small fisheries.

Ducks

Domestic ducks and other water-fowl that feed on water-plants can be used to control plant growth in ponds by grazing. However, their activities may also cause an increase in algal blooms.

Fish

Chinese grass carp will consume large quantities of aquatic plants in appropriate environmental conditions. Their rate of feeding increases with rising water temperature: at temperatures below 16°C their activities have very little effect, whereas food consumption is greatest at about 25°C. At this temperature they can consume their own weight in weed every day. It is probably unwise to stock with grass carp less than 18–20 cm long or 100 g in weight because these small fish are less efficient at eating plants. Brian Stott, a retired Inspector of Salmon Fisheries, for MAFF, has suggested that weed control in a pond could be achieved over a 2-year period with a stocking rate of 200 kg/ha of grass carp, assuming summer water temperatures exceed 16°C for prolonged periods. (Such ideal conditions are rarely achieved in Britain.)

Grass carp are selective feeders, preferring the 'soft' plants such as stonewort and Canadian pondweed to the more fibrous ones such as water lilies and emergent water plants. Consequently, in some circumstances they may not control unwanted plants, but instead overgraze those that are needed. It should not be forgotten that a pond that was initially stocked with the correct weight of grass carp may soon become overstocked as the fish grow. The subsequent loss of most or all of the valuable plant species may have a significant and long-lasting effect on the pond ecology. To effect controlled removal of excess weed, it may be necessary to remove some of these fish.

Efficient control may be possible if grass carp are used in conjunction with one of the more orthodox methods of control. It should be noted that there are legal restrictions on the introduction of grass carp into waters. Consent has to be obtained from MAFF to stock these species; contact the NRA for details.

If introduced into a pond in the correct density, carp will uproot the softer-stemmed plants whilst feeding in the pond mud. The increased turbidity that they create may exclude light and thus restrict or control plant growth.

Grass carp are used in Holland to control aquatic plant growth in dykes and other drainage systems. They are confined to certain dykes by the use of one-way fish debris weirs (see Fig. 2.13).

Fig. 2.13 Dutch fish debris gate.

Fertilization

The development of planktonic algal blooms which reduce light to a level that prevents the growth of rooted water plants has been used with success in America and in the British Isles. The pool is fertilized with soluble phosphate (e.g. 100 kg triple superphosphate per hectare) in the early spring as the rooted vegetation begins to develop. Until the correct balance is achieved, unwanted filamentous algae may sometimes develop before the macrophytes develop fully. There is also a real danger of creating a bloom of blue-green algae!

Trees

Trees overhanging a water (Fig. 2.14) can reduce light and so reduce aquatic plant growth. This shading effect is particularly noticeable beneath trees planted on the southern banks of rivers and lakes.

Polythene sheeting

It is also possible to create localized weed-free fishing areas in ponds by sinking weighted black polythene sheeting. This may be left *in situ*, or removed prior to fishing.

Herbicides

There are several different types of herbicide available for use in water, and they have several distinct modes of action. Herbicides that kill the parts of the plant with which they come into contact are called contact herbicides.

Fig. 2.14 Trees lining, and shading, a watercourse can reduce aquatic plant growth.

Herbicides that do not kill the plant rapidly in this way, but enter the plant itself, are known as translocated herbicides. As a general rule, only this latter group are any use in controlling re-growth or perennial water-plants. Herbicides can be further divided into non-selective types which will kill all plants, and selective forms which kill only certain species.

Before using any type of herbicide, it is important to identify the target plants on which they are to act. Accurate calculations must be made of the volume of water to be treated, using accurate data on water depth and area. Herbicides can be persistent and may prevent plant re-growth for many years, so great care must be exercised to prevent the unintentional death of all the aquatic plants present. If total eradication of plants takes place, the first species to recolonize are often the most undesirable in a fishery, like duckweed and filamentous green algae.

Small areas of the pond should be treated at a time. The main danger to fish life is not the toxicity of the herbicide, but the deoxygenation of the water following the death of the plants and their subsequent decomposition. Applications should be made before the plants are fully established since there will then be less vegetation to rot down; late spring is often an ideal period. On no account should a heavily weeded pond be overtreated in mid-summer as it is in these circumstances that fish are at greatest risk from deoxygenation.

The Control of Pesticide Regulations 1986 and MAFF Guidelines for the use of Herbicides in or near Watercourses and Lakes state that the NRA must be consulted before any herbicide is used. Incorrectly used, any herbicide could be harmful to humans, wildlife and fish and it is an offence to exceed the stated dose. Always follow the manufacturer's instructions on the label. A COSHH (Control of Substances Hazardous to Health) assessment may be required before the herbicide is used. It is advisable to hire a contractor trained and certificated in herbicide use to undertake any large scale plant treatment. See Appendix 6 for suitable herbicides for each aquatic plant.

2.6 Habitat improvement in still and running waters

The fish species present in still and running waters are influenced by natural migration, transfer by animals, or transfers by man (both deliberate and accidental). The number of fish present and their size are determined by the physical, biological and chemical conditions of the fishery.

Carp, for instance, might grow to a large size, and breed successfully, in a shallow lowland pool that would not support brown trout. A large deep upland reservoir, on the other hand, may produce large trout but only a few small carp because there is less food. Whilst it is important to stock waters with the species that are capable of flourishing in them, it is also possible to alter or improve the habitat to create better conditions for the fish. Habitat improvement methods are not recent innovations, but have been neglected recently in favour of that much over-used management technique–stocking (often over-stocking). Habitat improvement, correctly applied, often provides such excellent conditions for fish that destocking is necessary, and the removal and possible subsequent sale of these excess fish may provide sufficient money to fund further improvements. Although some of the available methods can be applied to both still and running water, these different habitats normally require separate techniques.

Physical improvement of stillwaters

The improvement of the physical habitat of stillwaters has received much attention in the USA and is gaining popularity in European fisheries, especially in Austria, Switzerland and South Germany. It is particularly suitable in large deep waters that lack underwater features, because physical improvements provide sanctuaries of cover and shelter, spawning sites and areas of improved fish food production.

Brushwood reefs can be made by tying bushes together and weighting them so that they sink. More recently, reefs of old car and lorry tyres have been used effectively for the same purpose. These are inert, do not rust or corrode, and will not contaminate the water. It is suggested that these reefs should occupy no more than 0.25% of the total area of the lake, and that they be split into several separate units.

Old tyres are often available from garages at no cost – providing they are collected. Each tyre will require several 2.5 cm holes drilled in the upper tread to allow air to escape, and sufficient bricks wedged in place on the lower side to sink it. Reef designs are a matter of individual preference, but some are shown in Fig. 2.15. Polypropylene or nylon cord or rope is used to lash the structures together, and the whole unit is lowered into position from a boat. These reefs require little or no maintenance. They provide not only an increased source of invertebrates, especially *Gammarus*, for fish but shelter from predation.

Chemical improvement of stillwaters

The chemical nature of the water in a fishery is important as it often determines the biological productivity of that water. Acid water is normally less productive than alkaline water. It lacks certain essential chemical constituents and, in particular, is poorly supplied with calcium carbonate. These chemicals are essential for the healthy growth of many plants, and acid waters rarely support dense populations of the aquatic creatures that form the principal food of fish. Most coarse fishers are at their most productive when they are slightly alkaline, with a pH of between 7.5 and 8.5.

It is possible to improve chemical conditions in new pools, and to reinstate

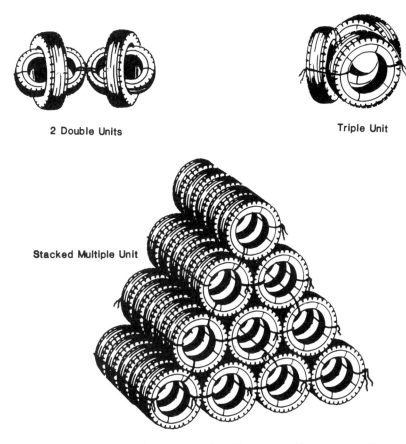

2 Double Units

Triple Unit

Stacked Multiple Unit

Fig. 2.15 Simple tyre reef designs, used as fish 'attractants' on the bottom of featureless ponds.

the productivity of old pools, by various means that fall under the general heading of fertilization. Old established pools decline in productivity with time because the nutrients necessary for plant growth become trapped deep in the pond mud. The pond soil turns increasingly acid, and the breakdown of organic matter like plants and leaves becomes slower. Fewer salts are released into the water because they become absorbed onto colloids in acid conditions and, although they are present in the pond soil, they are not available to plants.

In order to increase the productivity of an established pool, it is often necessary to raise the pH or alkalinity of the pond mud to promote plant growth. This is generally carried out by using lime. Lime is also important in another way, since it can provide reserves of the gas carbon dioxide which is necessary for the photosynthesis of green plants. New pools created in rich soil are normally quite productive, but pools such as gravel pits, clay pits, and sand pits are often nutrient-poor. In this situation, nutrients such as calcium need to be added, and these help to promote the formation of organically-based bottom sediments which consolidate the fine silt bed. Other nutrients important for plant growth (such as phosphates) may be in limited supply.

Fertilizers which can be used include hydrated lime which produces an alkaline reaction in the pond soil and aids the release of nutrients for uptake by plants. Lime should be spread over the pond bottom or water surface at a rate of 200–750 kg/ha, and should be used in waters where there is a rich mud bottom. If fish are present in the pool, care should be taken to treat only a portion of the water surface and apply the lime at weekly intervals. Crushed limestone should be added to nutrient-poor ponds where there is little pond mud and a lack of calcium. The dosage rate is approximately 750–1000 kg/ha. Basic slag is a slow-release inorganic phosphate fertilizer that promotes plant growth. It should be added in the winter, when a dosage rate of 300 kg/ha is normally recommended. Triple superphosphate is a soluble phosphate compound that quickly encourages algal blooms, which are in turn fed on by copepods and other crustacea. Although it may be expensive, it need only be applied at about 100 kg/ha.

Note that the timing of fertilizer addition is often important. Hydrated lime and crushed limestone should be added in the winter. Limestone can be added together with basic slag, but hydrated lime and triple superphosphate should not be added together since they react to form an insoluble compound.

Other fertilizers include well rotted farmyard manure, which can be added at a rate of about 1 kg/m^2. Great care must be used since the decomposition of organic manures in water can cause extremely low oxygen levels which may kill fish. Sewage sludges may also be used, but the same deoxygenation effects apply. If these are used, care must be taken to ensure that they do not contain toxic substances. If it is practical, and legally permissible, pools should be dewatered over winter so that the mud becomes oxidized on exposure to air, enriching the pool. When the lake is refilled, there will be a release of nutrient salts to the water, which will increase productivity.

Occasionally, enclosed and running waters may contain excess nutrients, and this enrichment can promote excessive plant growth, thereby lowering summer dissolved oxygen levels, especially at night. The alleviation of the problem is best achieved by reducing the nutrient inflow, but dredging or rendering the nutrients inactive by chemical means can also be successful.

Biological improvement of stillwaters

The provision of reefs to increase fish numbers, and the fertilization of still-waters to feed them, may produce a measurable improvement in the size and abundance of the fish present. It is also possible to assist this habitat improvement process by the deliberate addition of freshwater plants and animals.

It must be emphasized that although many 'stockings' of plants, fish food or fish are successful, many more are not. If the chemical and physical properties of the water do not suit a particular species, it will become extinct from the fishery sooner or later. Freshwater shrimps or snails, for example, will not survive in soft water.

Plant material and availability

Figure 2.16 shows a cross-section of a pond showing the different zones occupied by plants. Similar zones exist in rivers and lakes.

Generally, the vegetative parts rather than the seeds of plants are used for stocking. The roots or rhizomes can be purchased from a commercial nursery or aquarium, or transplanted from the wild. It must be pointed out that it is an offence to uproot any wild plant without the permission of the land-owners, and in some cases even with permission (Wildlife and Countryside Act 1981). It is sometimes possible to obtain stock plants from a nearby lake or pond, but in no circumstances should rare species of plants be dug up without the owner's permission. Where possible, only locally obtained or common plants should be used. Increasingly in recent years various growers have started to produce some species from seed. The County Wildlife Trust will also offer advice.

Suitable plants

All ponds will require plants both around the banks and in the water. Aquatic plants can be divided into four categories (see Fig. 2.16 and Section 2.5):

- For the water's edge, watercress can be established by pushing cuttings into mud and anchoring them with stones. Marginal reeds and rushes may be planted in a similar manner.
- For water up to 0.5 m deep, water starwort, water milfoil and water buttercup are suitable.
- For water 0.5–1.25 m deep, broad-leaved pond weed and lilies of various types are suitable. Only native lilies should be used. The establishment of lilies will be accelerated if they are planted in sacks or cardboard tubs of 50:50 rotted manure and soil.
- For deeper water, stonewort is an excellent choice. There are, however, several plants to be avoided, like Canadian pondweed, amphibious bistort and fringed water lily, which can quickly choke a pool, and for which there is no effective chemical control.

Planting

Submerged and floating leafed plants are generally planted by attaching a weight to the rhizomes and throwing them or dropping them into the water at the required place. In cases where there is very little organic matter on the bed of the lake, as with newly constructed lakes, it is advisable to put the rhizomes

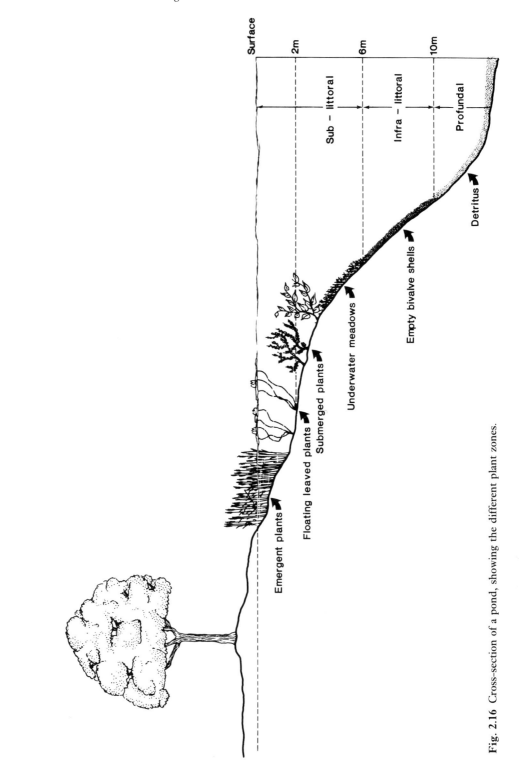

Fig. 2.16 Cross-section of a pond, showing the different plant zones.

in a biodegradable container, such as a sack containing soil and compost. Emergent species should be notch-planted in the shallow margins.

Animals

The most valuable introductions are often snails, pea and swan mussels, and freshwater shrimps. Shrimps should be released in shallow, hard water where the bottom is sandy and there are waterplant beds; snails are best 'sown' in shallow water where the bottom consists of stones.

Habitat improvement in running waters

Habitat improvement techniques in running water are confined to those of a physical nature; chemicals are quickly flushed out of a river (besides being illegal – in Britain – without the Regional NRA's consent), and plants and animals can migrate if conditions are not suitable. There is one notable exception to this – namely the use of lime to ameliorate the effects of acid rain (see Fig. 2.4).

'Instream' river improvement devices include those that impound or modify river flow (current deflectors, low dams and weirs, bank stabilization devices, etc); devices that provide direct cover (submerged shelters, artificial bank cover devices); and those that improve spawning areas.

Current deflectors (gabions, groynes and wing deflectors)

These structures utilize the natural river flow to create pools and riffles, increase water speed, and direct water flow. Some examples are shown in Figs. 2.17 to 2.19.

Low dams

Low dams or weirs are the most commonly used river improvement devices. Dams raise the water level above them and may provide more shelter for fish; after a while siltation may occur to such an extent that the original benefit of shelter is lost, but until this happens the settlement of silt behind small dams improves water quality. Below them, the water depth is reduced, and water speed increases, with scour holes forming. Bank erosion can be prevented by using one of the techniques discussed below.

Low dams are effective in recreating the pool and riffle nature of a stream. Typically, riffles occur at distances of between five and seven channel widths, apart, down the stream. Where this natural pattern has been destroyed (perhaps by a land drainage scheme), it may be wise to incorporate deflectors and dams at intervals of this length, the exact spacing depending upon the gradient of the stream. It is here that a thorough survey of the river flow and understanding of erosion and deposition is needed.

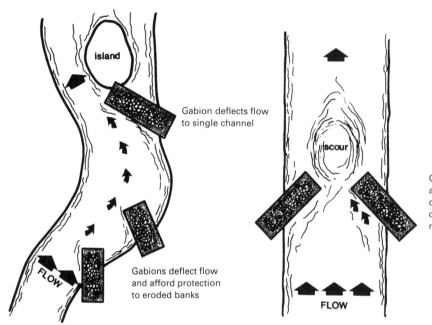

island

Gabion deflects flow
to single channel

scour

Gabions sited to direct flow
and result in the formation
of a pool through scouring
of the stream bed in shallow
riffle areas of river

Gabions deflect flow
and afford protection
to eroded banks

FLOW

FLOW

Gabions (wire baskets filled with stone), may be sited to counter or deliberately create erosion of a river's banks or bed.

Fig. 2.17 Some examples of current deflectors used in rivers to direct water flow for control of erosion.

Fig. 2.18 Pair of stone-built deflectors on the River Smite, Notts – built to scour out trout-holding holes in the downstream position.

Fig. 2.19 Pair of upstream-facing deflectors on the River Perry, Shropshire.

Bank stabilization devices

Many factors may increase the erosion of banks, and this is often a significant contributor to an increased sediment load in rivers. Bank stabilization techniques include stoning (Fig. 2.20), piling and the fencing of areas likely to be affected by cattle or sheep. This type of 'thrown stone' can be particularly beneficial to small species such as loach and bullheads and thus provide food for larger predators. Willow cuttings have also been used for stabilizing eroding banks.

Devices that provide direct cover

Shelter is important in order to provide cover for prey and predator; it also increases habitat diversity and provides spawning areas and shading. Cover may encourage the colonization of invertebrates, and help fish to establish and recognize their territory.

Artificial cover devices

These are designed to serve the same function as overhanging banks, hedges or bank vegetation. The devices consist of platforms constructed above the water surface and held in position by piles driven into the bed (Fig. 2.21). Floating overhanging platforms are an alternative solution. Innovative designs

Fig. 2.20 Bank erosion on this section of the River Trent has been stabilized by stoning.

made from concrete ferrocement or polymers are becoming available for such work.

Submerged shelters

Submerged shelters include branches, bushes or logs anchored to the river-bed, or submerged boulders or rocks. Such devices should provide minimum resistance to flow and should be designed to minimize the risk of trapping debris; for example, they should be placed in line with the flow, not across it. (Any in channel work may require a land drainage consent from NRA. It is advisable to discuss any planned work with them first.)

Spawning area improvements

Most salmonid species, and some cyprinids such as dace and barbel, spawn on gravel where the substrate size and the water speed and depth are suitable. It is possible to provide areas of gravel for spawning, but this is normally very costly. Gravel beds can also be raked over every year to remove silt. It is more feasible to use some of the previously mentioned devices (such as deflectors and dams) to create local eroding conditions to remove silt, so exposing suitable gravel deposits. It should be stressed that these devices should never be built without prior consultation with the appropriate NRA Region or Internal Drainage Board (IDB).

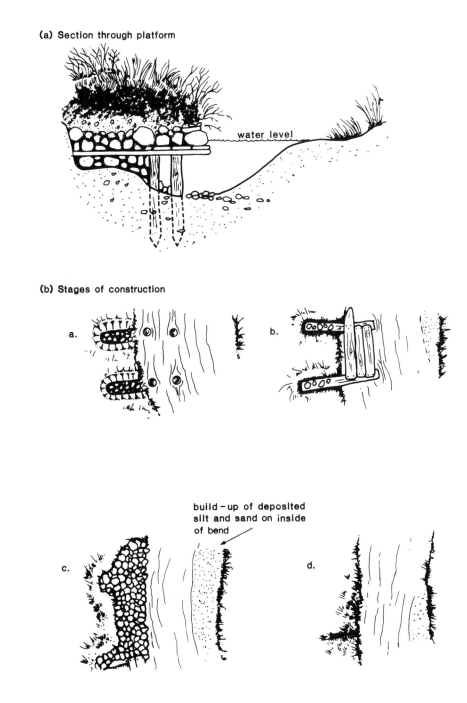

(a) Section through platform

water level

(b) Stages of construction

a.

b.

build-up of deposited
silt and sand on inside
of bend

c.

d.

Fig. 2.21 How a stream can be improved by constructing an emergent platform.

2.7 Bankside vegetation

Trees are an essential part of the environment and are hosts to a large number of organisms – lichens, fungi, worms, liverworts, insects and other invertebrates, as well as mammals and birds.

Advantages and disadvantages of trees

Trees provide windbreaks, create cover for fish, and their small feathery roots are often used as a substrate for spawning. These roots may also provide cover for fish, while the abundant insect life which develops in the tree canopy augments the pond and river food supplies. The shade created by trees can also reduce weed development in pools and rivers. They have a valuable function in stabilizing the banks, preventing erosion of meandering river courses. Generally, the planting of trees on the bank of an otherwise barren watercourse will improve the fish-holding capacity of that water, but careful attention must be given to siting.

On the other hand, too many dead leaves settling and covering the bottom of lakes and ponds can smother underwater plants. Again, once they reach full height, trees bordering rivers create so much shade that no weed grows beneath them. The current is then drawn to the weedless area and bank erosion follows (Fig. 2.22). They can also fall down, creating drainage problems.

Fig. 2.22 Bank erosion can be caused by water being drawn to the weedless area created by excessive tree shade.

Planting

Detailed guidance on how to plant and take care of trees is available from many sources. A good guide is contained in the Forestry Commission's leaflet *Trees and People*. Further advice is available from English Nature, the Forestry Authority, County Wildlife Trust and MAFF. County Councils administer grants for tree planting for the Countryside Commission.

It is preferable to plant trees in clumps rather than at uniform intervals; apart from the aesthetic appearance, this allows access for anglers. It may be possible in certain circumstances to receive a grant for tree planting (see Appendix 4).

Types of tree suitable for the waterside

There are two main groups of trees that occur on the waterside: the alder and the willow families. Some trees, like oak and ash, prefer drier habitats, but sometimes grow on the waterside as isolated trees or in woods. If any removal of trees from the waterside is envisaged the less easily managed species, like oaks, should be retained. The manageable trees are treated in different ways.

Alder and birch

Alder is probably the most common waterside tree. From the fisheries aspect it ranks highly for the number of insects that inhabit its foliage and ultimately fall into the water, thus acting as a food source for fish. It can also be cut to the ground and allowed to shoot again. New alders (plants about 1 m in height) should be protected from browsing animals for about ten years, by which time the leading shoots will be well out of reach and the bark not easily damaged. Silver birch is often planted, but the hairy birch is preferable as it grows better in wet soils.

Willows

Willows can be managed by pollarding or coppicing. The latter cuts the trees to the base and the former leaves a bole about 2.5 m high, where new shoots are out of reach of cattle (see Figs. 2.23 and 2.24). Willow trees are easy to establish if straight branches are cut and rammed firmly into the ground at the water's edge. Their shoots will require trimming every ten years or so, providing material for other uses. The white willow and crack willow will both provide good pollards. Smaller bush willows can be propagated by cuttings. They should be cut back close to the ground if they grow too large, and this action will also promote bush growth.

Fig. 2.23 Pollarded willow.

Fig. 2.24 Uncut willow, showing the danger of overgrown branches blocking the channel.

2.8 Control of pests and predators

The aquatic world provides an attractive environment for many different types of animal, with a vast range of habitats under, on or near the water. A fisheries manager, in planning the development of a particular fishery, should have decided on his plan of operation at an early stage. This plan will give details of the species to be encouraged at that water. He will also know from the plan which other species of fish – or other animal – will be unwanted – a pest to his site or a predator on his stock.

A predator on one fishery may be the preserved and preferred specimen in another; pike, for example, may be unwanted on a specialist trout fishery, but preserved on a specimen pike fishery. It is therefore not possible to give an exact and definitive list of pests and predators. Essentially, there are three main types of animal (four if invertebrates are included) that fisheries managers might wish to control: birds, mammals and fishes. There are many instances where it makes more sense to manage predators and 'pests' as part of the fishery, rather than to go to the effort of undertaking expensive and time-consuming control measures.

Problems caused by predators

Four main problems may arise on a fishery due to the presence of pests and/or predators:

- Direct predation on the preferred species, either by the predator eating or damaging stock;
- Disease and parasites spread by some species of animal;
- In many cases, physical damage to banks;
- Alteration of water quality, if the 'unwanted' species is present in excessive numbers.

Before undertaking control measures, fishery managers should remember the value of predators to fish stocks and man. They can remove fish weakened by disease and parasites, they can remove other predators, and they can control stunted fish populations. Predation on young fish by older fish of the same species is also normally high.

Birds

There are over 200 species of bird that live in or by the water. The vast majority are either better suited to marsh or estuary, or are visitors making use of a temporary food source.

The species that depend on water include representatives from several families. Those that feed on aquatic plants and animals often remain throughout the year, but the insectivores have to migrate when summer food supplies are not available. The ability of birds to react rapidly in adversity (they fly away!) gives them a clear advantage over other animals and makes them early indicators when anything disrupts lower levels of the food-chain. Climatic changes, increases in pollution, or subtle variations in amenity use can affect birds in spectacular ways, yet their adaptability enables them to cope with all but the most serious problems.

Heron

This is an unmistakable bird of the waterside, although it nests some distance away in tall trees. Though the heron has a widespread distribution in Britain, it is relatively scarce: the breeding population in England and Wales is approximately 5000 pairs. The heron is carnivorous and captures small fish, tadpoles, frogs, small mammals, small birds, reptiles, molluscs and insects from shallow water. Peak feeding times are at dawn and dusk.

The daily food requirement of an adult heron is approximately 370 g, but when young are being fed this figure is approximately doubled. In any natural fishery this supply of food is readily available, and control measures are rarely necessary. However, losses on trout farms can be significant. Calculations on one particular farm showed that during the peak feeding period between April and August, 1818 kg of trout were eaten by herons. When losses reach this level, fish farmers should take active steps to deter the birds.

Caging the site is the complete answer, but the cost of this would be great on large farms. The effect of scaring devices such as scarecrows, gunfire and flashing lights lasts for only four or five days as the birds gradually get used to them. Other ways of reducing the fishing success of herons at fish farm pools are shown in Fig. 2.25.

A dog trained to chase herons away is also a useful, if unusual, control method. If all else fails, there is provision in the Wildlife and Countryside Act 1981 (WLCA 1981) for an authorized person to kill herons with a legal method if it can be shown that such action is necessary for the purpose of preventing serious damage to the fishery.

Gulls

Under this heading come such species as the herring gull, black-headed gull, lesser black-backed gull and greater black-backed gull. It is now fairly well established that all such birds can spend a large part of their lives away from the sea. All species will eat almost anything edible, from fish to carrion and gleanings on rubbish tips. The larger gulls will also take small birds, chicks, eggs, etc. They will, given the opportunity, prey directly on fish, favouring the sort of easy prey that is available on trout farms. They are an important link in the life-cycle of eye fluke and for this reason alone should be actively discouraged. Trout farms in Denmark are required by law to place wires over

a) Earth banks gradient <60°

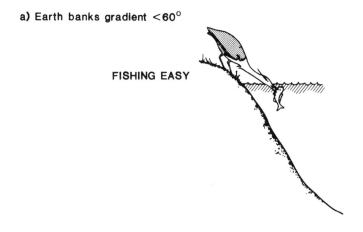

FISHING EASY

b) Twine and floats

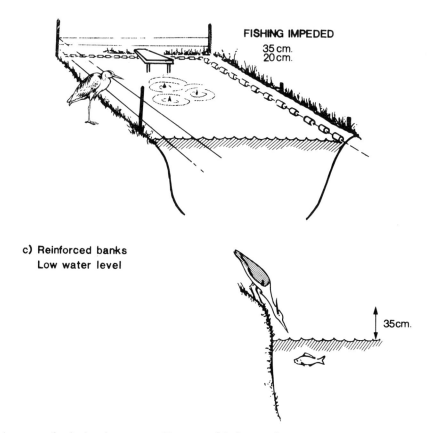

FISHING IMPEDED

35 cm.
20 cm.

c) Reinforced banks
Low water level

35cm.

Fig. 2.25 Other ways of reducing the success of herons at fish farm pools.

their ponds to keep these birds away. Any of the methods used to control herons can also be employed.

In Britain there is provision in the WLCA 1981 to kill herring gulls, lesser black-backed gulls and greater black-backed gulls by authorized persons; though it should be remembered that certain gull species are Schedule 1 protected birds.

Cormorants and sawbill ducks

There is some evidence that there has been an increase in cormorants and sawbill ducks (goosanders and mergansers) (see Fig. 2.26) in various parts of the UK.

The breeding cormorant (Fig. 2.27) population in Britain increased by about 5% during the years 1969 to 1987. The cormorant is normally a solitary feeder and will travel up to 50 km from its roost to feed. Research in Holland has shown that the daily consumption per bird of fish varies from 425 g to 700 g (15–17% of the bird's body weight). Although breeding colonies are primarily located in specific coastal areas, some inland sites have recently become established. The beginning of the increase in wintering numbers coincided with the cessation of shooting in 1981 (when cormorants received full protection status); this was also a time during which artificially stocked inland waters proliferated.

A recorded example of cormorant behavioural change has been noted by the Nottinghamshire Wildlife Trust. Overwintering (1985–86) counts at three Midland stillwater sites varied from 83 to 94. Recent observations show that the population has increased threefold since 1986. Significantly in 1991 there were 60 nesting pairs at these inland stillwater sites. It is speculated that the cormorants have changed their behaviour to match the concentrated quantity of fish.

Problems with cormorants appear to be largely restricted to inland still-waters, especially put-and-take fisheries. A few rivers also have cormorant problems.

Sawbills have increased in Britain and both species have extended their breeding range southwards. The trend has been particularly noticeable in Wales.

Cormorants and sawbills are granted full protection by the Wildlife and Countryside Act 1981, which implemented the EC Birds Directive. The issue or use of a licence to kill cormorants or sawbills at a fishery is a last resort measure. Moreover, due to the nature of the cormorant and sawbill populations, shooting is ineffective as a population control measure.

The NRA has no specific duties relating to either the protection of wild birds or the issuing of licences to control predation. Applications for control are referred to MAFF or the Welsh Office Agriculture Department (WOAD).

Eider (Somateria mollissima)

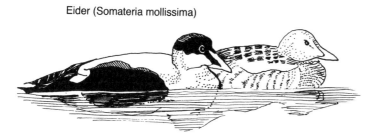

Kingfisher (Alcedo atthis)

Jay (Garrulus glandarius)

Goosander (Mergus merganser)

Cormorant
(Phalacrocoroix carbo)

Fig. 2.26 Sketches of various birds.

Fig. 2.27 Cormorant at rest on inland stillwater.

Ducks

Some species of duck such as tufted duck are thought to prey on small fish. This duck dives to depths of 2–3 m for its food – normally crustaceans, molluscs, insects and their larvae, as well as vegetable matter. If ducks are allowed to congregate in numbers on small pools, their droppings can affect the water quality. There is also a risk of passing on parasites to the fish in the water. Ducks can be controlled by most of the methods employed to discourage herons. In Britain there is provision in the WLCA 1981 to kill certain species in the open season from August to February.

Grebes

The great crested grebe and lesser grebe or dabchick feed their young on small insects, molluscs, etc. The diet of the adult bird consists mainly of fish and insect larvae. Grebes very rarely visit the ponds on fish farms, and appear to prefer lakes and gravel pits. There are no records of control operations against this species, and it is not lawful to kill them.

Kingfisher

This bird nests in a burrow excavated in a steep bank or embankment, and a pair of birds can have two clutches between April and July, each of six or seven eggs. The kingfisher catches small fish up to 10 cm long, and also crustaceans and water insects. When hunting prey it dives into the water from a bankside perch. If the owner wants to discourage these birds on fish farms containing fry, the ponds should be covered with fine mesh. In Britain, kingfishers are protected by special penalties under Schedule 1 of the WLCA 1981.

Canada goose

The Canada goose (Fig. 2.28) is a victim of its own success. After being introduced to St James's Park, London in 1665 for ornament and sport, it multiplied steadily to around 10 000 birds by the 1950s. Two factors contributed significantly to a further huge increase in numbers, now standing at 60 000 and forecast by the Wildfowl and Wetlands Trust to double by the year 2000. First, wildfowlers stopped shooting them because they were not considered a sporting shot; and second some 2000 were moved from areas where they were eating crops and fowling pasture land, to places that had no Canada geese. The RSPB concedes that there is probably a need to cull the geese in some areas. The Canadian goose is beginning to displace native species. These geese tend to stick together in gangs and get aggressive towards other birds, such as mute swans, pecking them and driving them away. Some fisheries managers may find the incidental damage large flocks cause unattractive for their fisheries. It is suggested they contact the RSPB.

Fig. 2.28 Canada goose, seen here with other species.

Other birds

Many birds have been observed taking fry from fish pens, tanks and ponds, probably because the fry are easy to catch. They include moorhens, coots, blackbirds, crows, robins and wagtails. All that is needed to discourage them is to cover the tanks with netting or wire mesh.

Mammals

Mammals that are semi-aquatic and adapted to live near or on water usually have good fur. In addition, species like the otter are adapted to aquatic life with webbing between their toes and a strong muscular tail. The fisheries manager will also come across other mammals, such as the brown rat and mole which frequent the land near water.

Water vole

Water voles are territorial: males have territories of about 130 m^2 while females take up smaller areas. They restrict themselves for most of the time to the edges of waters, and make tunnels in the banks with exits above and below water level. They feed on the leaves, stems and roots of waterside plants such as reeds and grasses.

Voles are becoming increasingly scarce due to a lack of habitat. However, when enough of them are present, their burrows can cause erosion in banks. The animals may be trapped in cages baited with apple, lettuce or carrot. They can then be relocated to a suitable site.

Otter (Fig. 2.29)

Otters are now rare in all areas and absent from most. They are essentially animals of rivers and marshes. Male otters may travel up to 10 km per night in search of food – usually fish or other aquatic life. A male otter is longer than a fox (about 120 cm) and twice its weight: mink, which can be wrongly identified as otters, are only half this size, at 60 cm. In Britain, otters are fully protected under the WLCA 1981.

Mink (Fig. 2.29)

This animal originated in North America, and in Britain is bred in captivity for its fur pelt. The first record of escapees breeding successfully in Britain was in 1956, and mink have now spread to most counties of the British Isles. They are very efficient nocturnal predators, killing fish, poultry, moorhens and voles. Mink populations can live side by side with other animals providing the habitat is diverse enough to sustain a wide species variety. On degraded rivers mink may become a problem.

The animal can be killed legally; MAFF no longer offer advice on their control. The Mink and Coypu (Keeping) Order prohibits the keeping of these

Otter (Lutra lutra) c.1:8

Mink (Mustela vison) c.1:5

Fig. 2.29 Otter and mink to show size difference.

two animals in Great Britain except under licence. The legislation also requires occupiers who know of any unlicensed mink or coypu on their land to inform the appropriate agricultural department. Mink are relatively easy to control once their presence has been identified. Cage trapping is the method of choice. Shooting with a powerful air rifle is also effective. Notification has been received from some prolific fisheries that mink populations achieve a 'natural balance'.

Coypu
The coypu has become established in East Anglia (mainly in Norfolk, Suffolk and North Essex) following escapes from coypu farms in the 1930s and 1940s. Adult coypus are easily recognized by their large size; males reach an

average of nearly 7 kg and are almost 1 m long from nose to tail-tip. They are almost entirely vegetarian but will eat freshwater mussels. Their burrows are about 20 cm in diameter with a half-submerged entrance, and they penetrate for several metres away from the water, thus lowering the strength of flood banks.

A scheme to eradicate wild coypu was launched in 1981 by the then Anglian Water Authority and Internal Drainage Boards. The animals are caught alive in cage traps and then killed humanely.

Mole

This animal feeds almost exclusively on worms, and builds complex tunnel systems on different levels. The mounds thrown up along the tunnels may interfere with fishery bank maintenance operations such as grass cutting. In extreme cases it is possible that the tunnels may weaken banks.

Common rat (brown rat)

The common, or brown, rat is widely distributed (Fig. 2.30), and can carry fish diseases as well as diseases that affect man. Leptospirosis or Weil's disease is a bacterium, carried in the urine of rats, which if contracted by people can cause serious illness or death; 60–80% of the rat population is infected. Reports of the disease have increased in recent years. The bacterium enters the body through cuts or abrasions on the skin. Fisheries managers, anglers and all aquatic recreationalists should follow a few simple hygiene rules, i.e. ensure all cuts are covered immediately with waterproof plaster, thoroughly wash or shower after immersion and ensure ready access to waterproof plasters and washing facilities.

The rat is a very adaptable animal and, if it lives beside water, soon becomes expert at diving and swimming beneath the surface. Its burrows are generally

Fig. 2.30 Common or brown rat.

dug in banks or other raised ground, and are complicated branching systems with nest chambers at intervals. They sometimes extend for several metres at a depth up to 50 cm.

Breeding takes place at all times of the year with up to 16 young born at a time. Young rats are easy prey for predators such as owls, foxes, stoats, weasels and cats, and both foxes and cats will take full-grown rats.

Rats can be a major pest on a fish farm; for example, by fouling and damaging fish food bags, and damaging nets and ropes. The animals can be trapped, poisoned, or hunted with ferrets or dogs. Probably the most effective way to control, if not eradicate, this pest is to carry out a properly designed poisoning programme. However, it is better to prevent rat problems by proper management than to have to resort to poisoning. Proofing buildings against rats, although more expensive, is a far better method of control.

Fish

Unwanted fish stocks can be removed by any of the methods described in Section 1.6.

Invertebrates

Some of the more serious predators at hatcheries can be invertebrates such as water beetles, dragonfly nymphs, water boatmen and water bugs. Some of these may cause enormous losses to young fish if not controlled.

2.9 Protection of fish stocks by regulations

Fishery managers are warned that this section is written with the intention of pointing out techniques and avenues open to them and, generally, how to apply them. It is not intended to be, nor should it in any sense be used as, a guide for legal purposes; however, where applicable, reference is made to the relevant Act and Section so that a subject can be pursued, if necessary, via that route.

Most of the text so far has been devoted to establishing viable fisheries by examining the various means that are available for their improvement and, assuming that this has been achieved, it is essential that it be maintained. To give effect to this we must consider the regulation of pressures on those stocks, through control of the users – and abusers!

Regulations

Fishery regulation in England and Wales is not an integrated service but operates at three different levels and involves elements of both the civil and criminal law.

Table 2.3 Summary of regulatory methods and their effects

Technique	Regulatory level		
	National, under 1975 Act	Local, under byelaw	Fishery, under local rules
Close seasons (angling)	Set under section 19 Salmon 31 Oct–1 Feb Trout 30 Sep–1 Mar Coarse 14 Mar–16 Jun (includes eels) Exceptions for specific purposes with NRA written consent	Amend national close seasons but duration not to be less than national. Can abolish close season for eels or set one for rainbow trout	Can impose own close season, provided it is longer than that set under byelaw
Minimum size limits	Only for salmon under Section 2	Can be set for any other fish	Can be set at any size, provided it is not less than that set under byelaw
Sanctuaries	None for angling Section 17 prohibits fishing near obstructions – does not apply to angling	Can impose restrictions as to time and area to prevent over-exploitation	Can impose any that are necessary to prevent overfishing
Limits	None for angling	Can limit numbers killed to help conserve stock	Can be set if necessary
Restriction of fishing effort	None for angling Section 26 can limit numbers of commercial implements	Usually controlled by close season byelaws	Can set a limit on number of rods that the fishery can accommodate
Restrictions on fishing tackle	Sections 1 and 5 prohibit many types of implement but permit the use of an unbarbed gaff, tailer or landing net as an ancillary to angling	Restriction on hook size, use of gaff, keep net size, etc.	Can impose any that are necessary in the interests of the fishery
Restrictions on other kinds of fishing gear	Section 5 prohibits use of poisons, explosives and electro-fishing gear. These can be permitted for specific purposes by the NRA	Usually apply to commercial fishing only	Not applicable
Stocking	Section 30 requires NRA consent for any stocking	Not applicable	Not applicable
Dealing with poachers	Schedule 1 Theft Act 1968	Not applicable	Can use the Theft Act or common law

National legislation

Overall control is exercised by central government, primarily through the NRA in respect of salmon and freshwater fish, and is contained in various Acts of Parliament and Ministerial Orders. The principal Act is the Salmon and Freshwater Fisheries Act 1975 (referred to as the 'Act' in this section) which has been amended and/or strengthened in all or part of its implementation by subsequent legislation including the Fisheries Act 1981, the Diseases of Fish Acts 1937 and 1983, the Salmon Act 1986 and the Water Resources Act 1991.

The function of the legislation is, *inter alia*, to provide protection for fish stocks by preventing illegal methods and overexploitation, creating obstruction-free passages for fish, regulating times of fishing and selling fish and providing powers to ensure that the provisions can be enforced. It also provides a framework within which more local regulations may apply.

The Act is divided into sections:

- *Section 1* prohibits the use of certain implements for catching fish including snatches, spears and set lines;
- *Section 2* prohibits the taking of immature (undersized) or unclean fish and the use of fish roe as bait;
- *Section 3* regulates the use of nets;
- *Section 4* illegalizes the releasing of poisonous substances into waters containing fish;
- *Section 5* prohibits the use of poisons, explosives or electrical devices for catching fish;
- *Sections 6–18* deal with obstructions in water courses frequented by migratory fish;
- *Sections 19–24* deal with the close seasons and times during which fish may not be legally sold – these are further elaborated in the Schedules of the Act;
- *Sections 25–27* deal with fishing licences;
- *Sections 28–30* deal with the general powers and duties of the NRA; and
- *Sections 31–36* deal with the powers of water bailiffs.

The Schedules of the Act deal in further detail with matters contained in the various sections; thus Schedule 1 is concerned with close seasons, Schedule 2 deals with licences and Schedule 3 covers administration, including the power to make byelaws.

Byelaws

The second level of control is provided under Schedule 3 of the Act which allows the NRA to make byelaws that are intended to take account of local conditions which may apply in different parts of its area.

Currently byelaws are promoted by regions of the NRA and can apply to the whole or part of a regional area and to the whole or part of the year. The fact that there are ten regions of the NRA, and that byelaws are primarily of local application, means that byelaws can vary considerably from region to region. This makes it imperative that every fishery manager is in possession of a copy of the byelaws which apply to the region in which his fishery is situated.

The NRA byelaws often vary the close season for salmon and trout and sometimes that for freshwater fish. In some regions they permit the fishing for eels during the close season for freshwater fish. Fish which are exploited as food – the game fish, grayling, etc. – are frequently subject to greater byelaw control than those species which are returned to the water after capture. It is worth reflecting that there is no statutory obligation for an angler to return his fish to the river unless it is below the minimum size limit set under byelaw or caught accidentally during the close season for the species concerned; consequently, a fishery manager may consider that there is a need to exercise control over this by making a rule for his fishery.

Byelaws usually have the effect of making the provisions of the Act somewhat more restrictive than would otherwise be the case.

The enforcement of both the Act and the byelaws rests with the NRA through its bailiffing service. Powers to help implement enforcement are contained in the 1975 Act and subsequent legislation.

Fishery rules

The third tier of regulation rests with the occupier of a fishery through the introduction of rules which imposes a code of conduct upon those who visit the fishery legally and the use of the law to deal with those who visit it illegally.

Other rules may not relate directly to fishing but can cover such matters as parking, use of radios, disposal of litter, etc.

Generally, any contravention of the fishery rules can be dealt with under the civil law. Any person found fishing without permission can be removed, using reasonable force, or dealt with under the law relating to trespass, but this is a civil offence and some proof of consequent damage is often required if redress is to be gained.

Techniques

Within the levels of control discussed above there are a number of techniques available which can be used to protect fish stocks and prevent over-fishing (see Fig. 2.31). These include: the imposition of close seasons; the creation of closed areas (sanctuaries); catch (bag) limits; limitations on the numbers and use of various types of gear; and size limits. All these are used in one form or another on fisheries within the British Isles.

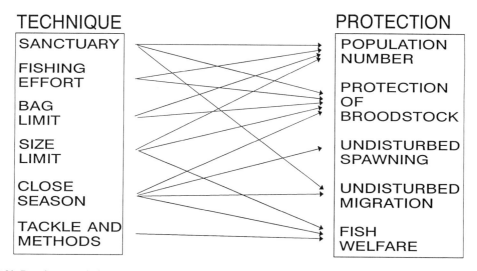

Fig. 2.31 Regulatory techniques used for the protection of fish stocks and the fishery requirements which are addressed.

Close seasons (the Act, Section 19)

The principal purpose of the close season is to provide a period of respite to enable fish to spawn unhindered and to recover from the effects of spawning.

Under the 1975 Act, minimum close seasons are mandatory in England and Wales for salmon, trout and freshwater fish but there are exceptions that can apply to freshwater fish and eels. In this context eels are considered to be freshwater fish although under local byelaw they can be *exempted* from the close season. Conversely, and also in this context, rainbow trout are not classified as trout and hence no close season applies – one may be imposed by a byelaw, however.

In addition to a close season there is also a close period each week during which it is illegal to fish for or take salmon or trout by any means other than rod and line.

The dates laid down in the Act apply generally and represent the core periods during which spawning takes place on a national basis. However, there are local variations in spawning times which fall totally or in part outside these. To allow for this byelaws can be introduced by the NRA to vary the *dates* laid down in the Act between which the close seasons apply, but any such byelaw must not effectively reduce the minimum *period* stipulated. The dates covered under byelaw can vary considerably between regions of the NRA.

The position was greatly strengthened under Schedule 1 of the Theft Act 1968 which made it an offence for any person to fish in private waters without permission and provided the fishery occupier with a means of dealing with the 'poacher' under criminal law. As a result an offender or would-be offender is far more easily dealt with: the accused need not have caught any fish – his mere presence on the fishery with intent to do so is sufficient. Furthermore, under this Act it has been held that a person fishing in these waters who is specifically in breach of any conditions under which he obtained his permission to do so, has rendered that permission invalid and is therefore breaking the law. This Act allows for the seizure of tackle and the arrest of a person fishing without permission but the latter does not apply to angling in the daytime.

Generally, fishery rules cannot be enforced by officers of the NRA; nevertheless these officers have the right of access under the Water Resources Act 1991 to cross land to enforce the Acts and byelaws, and can take action for any infringement of these on a fishery.

The making of fishery rules should not be rushed into. The problem often encountered, especially in respect of angling club rules, is the initial introduction of a large number of rules relating both to the fishing and the constitution of the club. In cases where a rule urgently needs modifying, unnecessary delay can arise, particularly if the approval of an annual general meeting is required to amend it.

The enforcement of fishery rules rests with the fishery occupier, either directly or through any keeper or agent appointed by him. Any person purporting to enforce the rules of a fishery on behalf of the occupier should carry with him a written authority to that effect. Furthermore, he must be instructed in his duties and made aware of his powers to act. For his part the occupier must be prepared to support his agent in any legitimate action he takes on his behalf.

There are certain exceptions set out in the Act which permit fishing in the close season; the more important of these are:

- For the purpose of artificial propagation of fish, or the stocking or restocking of waters, or for some scientific purpose, with the prior permission in writing of the NRA;
- For the removal, by the occupier, of a fishery where salmon or trout are specially preserved, of eels, freshwater fish or rainbow trout not so preserved; *and* fishing with rod and line for freshwater fish or eels in such a fishery with the permission in writing of its occupier;
- Fishing for eels where such fishing is authorized under byelaw;
- Fishing for freshwater fish or rainbow trout for a scientific purpose;
- Taking freshwater fish for bait with permission in writing of the occupier.

In addition to the close seasons laid down under the Act and byelaw a fishery occupier can impose his own in respect of one or more species within the fishery, *provided* that its duration is greater than that set under the Act or byelaws, and it must include the period which applies locally.

Size limits (the Act, Section 2)
Size limits relate to the minimum size of fish which can be caught and removed from a fishery – fish below this size are considered to be immature.

Size limits are introduced for one of two reasons:

(1) To allow each species to reach a size which ensures that they breed at least once, and
(2) To prevent the wasteful removal of fish not yet large enough to provide optimum sport.

The size at which a species may breed is related to its environment and the calculation of the best size limit will be a matter for each fishery to decide. This has to be carefully arrived at as, for example, if all or the majority of the breeding stock is removed after it has had a chance to breed only once, then the size of eggs that will be laid will be at the smaller end of the natural range since this largely depends upon the size of the female fish. In turn this can lead to reduced fry viability. To prevent this happening it may be advisable to impose a fairly large size limit. Conversely, on a very lightly fished water size limit becomes of less importance.

The Act makes it an offence for a person to take or kill any fish which is 'immature', which it defines as 'a salmon of a length less than 12 in, measured from the tip of the snout to the fork of the tail, and any other fish of a length less than that set by byelaw', but there is no offence if such a fish is caught accidentally and returned to the water immediately with the minimum of injury.

A fishery occupier can impose any minimum size limit he wishes in relation to any species of fish in the fishery *provided* that it is not less than the corresponding minimum size limit laid down under byelaw.

Sanctuaries
Sanctuaries are areas set aside in which fishing is not permitted, with the object of providing a haven for fish and thereby placing less demand upon a fishery as a whole.

Generally speaking the provisions of the Act do not allow for the creation of sanctuaries but these can be introduced by the NRA under byelaw. An example of this is found on some migratory fish rivers where there is a ban on fishing immediately above the tidal limit during the early part of the season in order to protect migrating salmon and sea-trout smolts which congregate in these areas. Certain areas in the sea around the mouths of rivers used by migratory fish and sections of river downstream of weirs are often treated in the same way to prevent the over-exploitation of salmon and sea-trout which congregate there before moving up river.

On private fisheries the creation of sanctuaries where angling is not allowed is useful and effective – particularly in certain trout fisheries where they help reduce the impact of angling on stock fish and effectively even out the short-duration surges in catches that would otherwise follow the introduction of stock fish. However, on enclosed waters even a large sanctuary area will not greatly reduce the numbers of trout that are eventually caught during the season, though it will help ensure a more even distribution of catch throughout. The location and size of such areas is at the discretion of the fishery occupier.

Baits and lures (the Act, Section 2)

The only bait restriction contained in the Act relates to fish roe, the use of which is prohibited.

Under byelaw, however, in the interests of preventing over-exploitation at local level, the use of certain baits or lures is prohibited or restricted. Some baits such as maggots are highly effective if used on a coarse fishery and have little effect upon stocks as the fish are subsequently returned to the water – but if used on a trout fishery, where the fish are removed, or on a river with large numbers of salmon and sea-trout parr, their effectiveness can lead to an unacceptably high loss of numbers. To this end bait restrictions are often found which run parallel with, and complement, the methods of fishing allowed.

The principle can be used on a private fishery in the interests of stock protection but the rules governing this must not run counter to anything in the Act or byelaws.

Bag limits

Bag limits, as such, are not covered by the Act although they may be imposed under byelaw in respect of certain species.

Private fisheries which expend a considerable amount of money in maintaining the quality of their fish at a high level almost invariably impose bag limits. These are generally applied, on a *per day*, or *per visit* basis, to fish which are caught and then removed from the water – such as on a trout fishery where the replacement of such fish to maintain a standard is both necessary and expensive.

Where fish are caught and then subsequently returned alive to the waters bag limits are not considered to be necessary although on fisheries containing specimen fish such limits may be deemed expedient.

Restriction of fishing effort

A restriction on the numbers of rods that may be used on a particular fishery serves the dual purpose of regulating the amount of fishing that is taking place, thereby spreading out the pressure on the fish, and also reducing the chance of interference by, or crowding of, anglers themselves. This can be achieved either by limiting the number of anglers permitted on a fishery at any one time or only allowing each angler to use no more than the prescribed number of rods (see below).

Restrictions on fishing tackle
Although under the Act restrictions and prohibitions are placed upon many types of fishing gear in general they do not apply to fishing tackle – except for the use of a gaff, tailer or landing net which is specifically permitted when used as an ancillary to fishing with a rod and line (the Act, Sections 1 and 25).

Where such restrictions exist they do so either under byelaw or fishery rules and relate in the main to limiting the numbers of rods, controlling the method of fishing and regulating the use of associated equipment.

Under byelaw the number of rods which are permitted can be limited but in most cases restrictions are imposed by fishery occupiers.

On some fisheries, depending upon the way in which they are managed, it may be necessary to limit the method of fishing (e.g. 'fly only'), to stipulate the minimum or maximum size of hook permitted, or to lay down a minimum breaking strain of line. This latter restriction is sometimes found on fisheries stocked with large trout or specimen fish of other species with the intention of avoiding breakages which result in fish escaping with the hook still embedded in their jaws.

The regulations relating to associated tackle can include the requirement that keep-nets are made of the correct type of material and conform to a minimum size and shape; this is to ensure that fish are not damaged during their retention in the nets. Many pike and game fisheries also prohibit the use of a gaff to help land fish.

Within the limits of rod and line fishing restrictions are usually directed towards avoiding the catching of, and resultant damage to, non-target species and ensuring that the type of fishery that is being managed is in fact providing the sport for which it was designed.

It should be noted that in some NRA regions the restrictions dealt with above may be covered under byelaw.

Restrictions on other types of fishing gear (the Act, Sections 1 and 5)
The management of a fishery may require fish to be removed from a water to be transferred elsewhere. There are several ways of doing this, all of which could be abused by someone wishing to take fish illegally, which of necessity requires the need for some form of control.

The use of nets is usually subject to the consent of the NRA, while using electro-fishing apparatus without consent is an offence under the Act.

To avoid falling foul of the legislation fishery occupiers should always consult with the NRA before using any method for removing fish other than by means of rod and line – and note that NRA consent is needed before any fish removed can be introduced elsewhere (see below).

Stocking (the Act, Section 30)
The introduction into a fishery of stock that originates from a source where

disease is rife can cause the disease to spread to the fishery's indigenous population and lead to heavy mortalities. The same can apply to waters infested by parasites. Similarly, the introduction of an exotic or alien species to a fishery, without prior thought being given to the possible effects that it might have, can also prove disastrous.

These are major problems which the Act attempts to prevent by making it an offence to introduce any fish into any waters without first getting the written consent of the NRA.

Fishery occupiers must be aware of this requirement and be vigilant to ensure that certain anglers do not succeed in introducing fish illegally into a fishery. This has happened on some trout fisheries where the introduction of coarse fish, without the knowledge or approval of the occupier, has had a detrimental effect.

Part 3
Exploitation

By this stage the fisheries manager will have a very clear idea of how the fishery should be used or exploited. This section explores the various ways in which this can be achieved, and also describes the different methods of aquaculture. Additionally, some indication is given of the needs of other recreational water users and how the business-minded fishery manager could use his fishery for other purposes.

3.1 Angling requirements and methods

In common with most recreational pursuits, the enjoyment that anglers derive from their sport is closely linked to quantifiable goals that they set themselves. These might include catching a particularly large fish of a certain species, winning an angling competition, or making a large catch of fish. It follows that the angler's basic need is for fish, be they large, numerous or of a certain species. There are, however, many other factors that contribute to making a fishing trip or a fishery enjoyable. Indeed, for some anglers a quiet picturesque fishery may be more important than the fish it contains.

General requirements

As most anglers travel to and from the fishery on a daily basis, it is important that car parking facilities and access to the water's edge are good. Obviously, a large clearly-marked-out car park with a firm base is preferable to a section of marshy ground on which cars are haphazardly parked or, worse, become stuck. Secure litter bins, emptied regularly, will help to prevent the parking area becoming untidy.

Approved routes to the fishery should be clearly labelled, and anglers should be able to cross any obstructions such as hedges and streams easily and safely. If access routes are poor or inadequately signposted, there is risk of damage to crops or fences. The banks themselves should be safe, and there should be a well-defined path leading to the fishing areas. Signs should be firmly secured. If they are to be placed on trees or buildings, it is wise to fix them higher than 3 m from the ground to help prevent them being removed or defaced by vandals.

Most fisheries are made more enjoyable if some measures are taken to provide firm, level and safe stations from which to fish. These may consist of simple, flat areas that have been cut into the bank, or of more elaborate platforms or fishing stages. If the fishery is popular or if angling competitions are held on it, it is worth numbering these fishing stations consecutively.

In Britain, angling is one of the most popular of all outdoor water sports. Among sea, game and coarse angling, the last predominates. Coarse or freshwater fishing can be further divided into pleasure angling and match angling. Many freshwater anglers take up the sport as a means of getting out into the open air, and the popularity of angling may be linked to its therapeutic value, or to the satisfaction that is gained by utilizing the hunting instinct which is said to lie in everyone. Whereas the pleasure angler pits his skills and wits against the fish, the match angler competes with other anglers by attempting to catch the heaviest weight of fish in a given time. The needs of coarse fishermen have been identified as follows:

'The basic, essential resource is an area of water well stocked with fish. The best conditions include a limited but varied bankside growth, some aquatic weed, and tree shade – particularly on the southern banks of rivers and streams. In well fished areas, a stretch of at least 7 m of bank for each angler is desirable. Strong or strengthened banks are required for safe access along rivers and streams, with occasional car parking facilities with toilets and refuse disposal facilities.'

Game fishermen require even more space and undisturbed water than other anglers. They require at least 20 m for backcasting so footpaths should be well back from the fishery edge. On rivers the ideal is a meandering channel with slow moving, deep areas, shallow riffles with a sandy or pebbly bed, cover for fish, and a wide variety of lighting conditions. Many British rivers and nearly all reservoirs are stocked to provide a supply of catchable-sized trout, although there are some natural self-sustaining populations. Salmon fisheries usually contain fish that are naturally bred, but a few rivers contain fish originating from eggs, fry or smolts stocked in preceding years.

Improvement of access facilities

There are several types of operation that a fisheries manager can undertake to improve access to a fishery. It is important, however, to check with the owner (if the fishery is leased), or with other interested parties (if the water is owned). Local neighbouring farmers, and persons controlling shooting rights, should be informed of any proposals that may affect their right of access. Planning permission may be required for any new access roads and the County Council highways should be consulted about new access turnings off main roads.

Fig. 3.1 A simple stile on the main footpath helps to prevent damage to fences and hedges.

Stiles
Properly constructed stiles prevent damage to fences, hedges and gates (Fig. 3.1). Single plank bridges (with supporting handrail), placed over small dykes will also prevent damage to the banks. The construction of a stile at one end of such a bridge will also control cattle.

Cattle grids
Normally, a farmer will install a grid in place of a gate where traffic volume would render the latter unacceptably inconvenient. In certain circumstances such as at the entrance to a busy car park (Fig. 3.2) it may be worthwhile constructing one, but this operation should be undertaken or supervised by qualified engineers.

Fencing
Post and rail or post and wire fences to control the movement of cattle should be constructed whenever possible. Cattle drinks should be lined with stone to control bank erosion, and fenced around the water's edge to allow the cattle to drink but not enter the watercourse. Fencing should not be constructed across the river as this will obstruct flow.

It is important to discuss the location and construction of fencing with the local farmer, and land drainage engineers of the NRA will also need to be consulted if the watercourse is maintained by the Authority. Drainage byelaws of some NRA Regions forbid anyone 'without the consent of the Authority to erect or place in the river any fence, stake, post, fishing rack, pen or enclosure for birds or fish'. It is always better to check first and so avoid the additional expense of removing the obstruction after it has been erected.

Fig. 3.2 A cattle grid entrance to an anglers' car park.

Angling platforms

The types of angling platform that are acceptable by many NRA engineers are shown in Fig. 3.3. Woods such as oak and elm (if available) may be utilized, and these should be treated to prevent premature rotting. Angling platforms should be constructed to allow water to drain back into the river, as the wood is likely to split and become unsafe if the platform retains water after floods.

Permanent pegs

Many angling organizations that have long leases and allow fishing competitions construct permanent 'pegs' which mark designated fishing stations. These are often short, numbered concrete, metal or wooden stakes. Consent from the relevant drainage engineer (as well as from the farmers and riparian owner) is often required if the banking is mowed by the NRA, as fixed pegs can interfere with grass mowing. Alternatively, concrete slabs of about 0.25 m^2, can be inset into the banking: this enables grass-cutting equipment to cut immediately over the 'peg', and provides a clean stable fishing area.

3.2 Formation and management of angling clubs

General considerations

Any body of persons with a mutual interest in a particular type of fishing or fishery that wishes to put its interests on a more formal basis can do so by

POSITION

The platform must be sited PARALLEL to the flow and project a MAXIMUM of 1000mm into the river, exceptions being on small narrow rivers with shallow banks or where navigation could be affected.

The platform must be no wider than 1500mm.

The platform must be no higher that 500mm above WATER LEVEL at normal flow.

Platforms MUST NOT disturb any bank revetment or protection work.

BANK CUTTING

Use tanalised board pegged and nailed in position to retain consolidated stone surface

The base should be seeded or stoned immediately after cutting to prevent scour.

The MAXIMUM dimensions for cuts are 1 metre into the bank and 3m along the bank.

The MAXIMUM depth of a cut should be 500mm deep from the existing line of the bank.

VERY STEEP BANKS

(where the bank needs retaining)
Such cuts must be retained with permissible material such as tanalised 100mm round timber, or webbed geotextiles, anchored by soil. The latter requires immediate seeding to encourage revegetation.

OPTIONS

Decking

Cut with Textile Reinforcement of vegetation

Hard surface using materials in keeping with the surroundings. Normally only suitable for urban areas.

SPOIL from cutting work MUST NOT be deposited in the river. It is to be REMOVED to above the bank top to a position where it does not affect the flood plain.

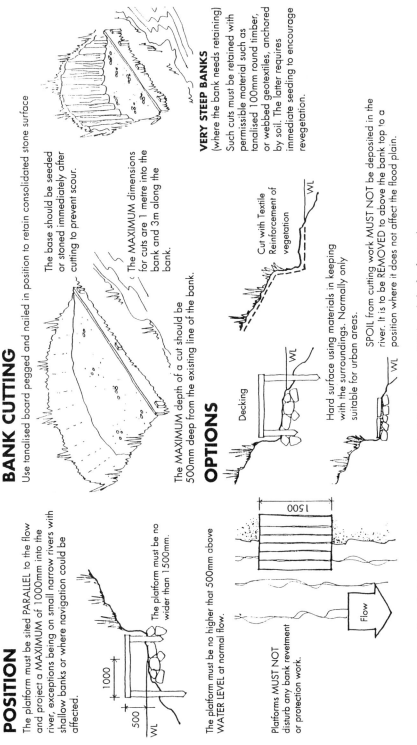

Fig. 3.3 Types of angling platform (note that consent is required from both the NRA and the landowner).

forming an angling club. Clubs can take many forms, and vary greatly in size from the ultra-large association, with a membership of thousands and many miles of fishing, to the small syndicate of friends who fish a single meadow.

The initial meeting

When there is a demand for the formation of a club the first essential is to call a meeting of all those interested and to have a simple agenda, e.g., the election of a chairman for the meeting, a motion that the club be formed, the election of a small working party to draw up rules and the election of a secretary.

This agenda is quite short and is intended to promote discussion. Too big an agenda at the first meeting can be counter-productive; it is often far better to let those attending give their views and suggestions within a very loose framework rather than confine them to a comprehensive agenda which might stifle or prevent the airing of some very good ideas.

Conducting business

Once a club has been established it will have to set up guidelines within which to conduct its business and these usually form part of the club rules (see below). Within the rules of most clubs there is provision for an annual or extraordinary general meeting, the creation of committees and sub-committees, establishing a quorum, determining voting rights, the election of a chairman and committee members, the appointment of officers, financial matters, etc.

The proceedings of all meetings should be recorded and minuted. This is usually the responsibility of the secretary but should he not be available one of the other members present should be delegated to take the minutes.

Rules

Whatever its size, a club, if it is to function efficiently, needs to have a set of rules to form the framework within which it conducts its business, and the members must agree to abide by them. Rules can be agreed verbally or formally incorporated in a document; the latter is especially desirable in the case of big clubs for which a printed set of rules to work to can prove an invaluable asset in conducting business and controlling members.

The rules can be divided into two broad categories:

(1) Those that deal with the business management of the club's affairs (sometimes known as the *club constitution*);
(2) Those that deal with the running of its fishery and the conduct expected of its members when fishing those waters.

If a club decides to split its rules in this way it will have to decide into which category a particular subject will be put.

It goes without saying that a small syndicate of friends will probably have the minimum number of rules, relying on one another's good common sense and judgement to run a very effective fishery, while a large club may require a considerable number in order to cover its wider interests.

A note of caution needs to be introduced here. An over-enthusiastic club can start off initially with a comprehensive constitution and an impressive set of rules relating to the fishery only to find at some later date that they are prevented from doing something urgently desirable because the constitution or the rules do not permit it. This is a situation to be avoided at all costs. A useful maxim is 'start off slowly'. Begin with the *minimum* number of rules that are required to get the club off the ground and add to, amend, or delete from these as necessary over the next few years until the right balance is achieved, but always leave enough flexibility for amendments to be introduced quickly if the need arises. Rules relating to the control of the anglers are discussed in this Section under 'Rules of the Fishery'.

The name

Every club needs to be identified and therefore needs a name. This should be coupled with an address to which all correspondence can be sent; usually this will be that of the secretary. Both the name of the club and the address from which it conducts its business should be included in any written rules and on any stationery used.

Aims and objectives

It is sometimes of great help if, on its formation, the reason for the club's existence and the things it sets out to achieve are included in the rules so that its members know exactly why and for what purpose the club was formed. Such may be obvious to founder members but in a few years time the original reasons may be overlooked, become blurred or be subject to mutation unless they are set down in writing and can be referred to.

Membership

If there is to be no limit or restrictions on who can join the club this needs to be made clear in the rules. Similarly, if membership is to be restricted the qualifications of those entitled to apply for membership need to be defined. Qualifications can be based upon place of residence, employment, age, membership of some other body, etc.

Initially a club should have only one category of membership but after a time it may become apparent that others are necessary to meet different interests and circumstances that have arisen within the club. A case in point is the retirement from work or redundancy of many of its members in which case it might be thought appropriate to introduce a *retired members* category.

Other likely reasons for establishing special categories include an influx of young members; a change of interest among members, e.g. more taking up trout fishing; the number of incapacitated or disabled members; the number of wives and friends who help the club but do not fish; people who have given long service to the club; and people not in the club but who have helped it. These could be accommodated by creating *junior*, *game*, *disabled*, *social*, *life* and *honorary* membership respectively.

Subscriptions

Subscriptions are required to finance the running of the fisheries and to meet the costs of administering the club. At its inception a club will have no money and in order to allow it to operate and meet initial expenditure members may be asked to pay a 'Joining Fee', which would be a one-off payment, in addition to an agreed annual subscription. This could also apply to all future members when joining.

Different categories are usually associated with differential subscriptions and if these form a big proportion of club membership and are at a discount they could give rise to an increase in the subscriptions of ordinary members.

Subscription rates need to be carefully worked out; they must be sufficient to meet the budget expenditure yet not so high as to deter potential members from joining or cause existing members to resign.

Financial sources

Club subscriptions on their own very rarely enable a club to purchase a fishery and the club is therefore forced to look elsewhere for finance to achieve this. Sports Council grants are available for this purpose while local authorities or businesses may be prepared to offer assistance in cash or in kind. One condition that appears to be general is that the club itself must also be prepared to put its own money into the purchase of a fishery. The advice of a bank should always be sought before embarking on such a project.

Banking arrangements

A new club will need to open a bank account. To open an account requires a deposit and it is here that the benefit of having 'joining fees', especially if the club is newly formed, is obvious as these can be used to open the account. The bank will issue a cheque book and will also need an authorization from the club allowing certain nominated persons (usually the treasurer and the chairman or secretary) to sign cheques on its behalf and transact its business with the bank.

Sale of fishing permits

If one of the aims of the club is to provide fishing for the general public as well as for its members, or if it needs to raise additional income, it can do so by issuing fishing permits. These are usually on a daily basis and are, *pro rata*, more expensive than the cost to a member of the club, e.g. the club annual subscription may be £10 whereas a daily permit could cost £2. (If someone wants a permit for a longer period he should be encouraged to join the club, if vacancies exist.)

Once a decision is taken to issue permits an outlet or outlets for their sale must be found. One likely source is a local fishing tackle shop, but whoever takes on the job will require some form of recompense for the work involved. An agency fee of a percentage of the gross takings is the most usual way of dealing with the matter.

On some fisheries the club keepers sell permits to anglers who are already fishing. This practice is not recommended as unless the angler is actually seen by the keeper he will not buy a permit. A much more realistic approach is to insist that, provided permits are easily available locally, no one fishes unless he has purchased one beforehand.

It is important, as will be seen later, that anyone, to whom a permit is issued, is also given a copy of the club rules and that club members are conversant with them.

Rules of the fishery

These rules (see also Section 2.9) are needed to ensure that the interests of the club are protected, that all who fish do so fairly and that the fishery is not over-exploited. To achieve this the rules should require anyone fishing in England and Wales to be in possession of a valid NRA rod licence and to observe the Salmon and Freshwater Fisheries Act 1975 and byelaws made under it.

The club may add its own constraints. They can stipulate where members and permit holders can park vehicles and the routes to be used to cross land to reach the water, prohibit the lighting of fires and damaging of waterside vegetation, require litter to be taken home, set a bag limit on the number of fish which can be taken away in a day, ban the use of certain baits and/or methods of fishing, and limit the number of anglers allowed on a particular beat at any one time. The inclusion of a rule which states that the rules of the fishery must be observed at all times is essential.

Discipline

The club needs to have a policy relating to its dealings with members or permit holders who break the rules of the fishery. Three courses are open: a warning, banishment or prosecution.

Both the issuing of a warning and banishment can have repercussions as the decision has no legal backing. However, prosecution under the Theft Act uses the criminal law and usually has the most salutary effect on offenders. If the rules of the fishery are broken, or if someone is fishing without a permit, that person is fishing unlawfully and can be prosecuted under the Theft Act 1968. To overcome the excuse that the angler did not know the rules it is essential to issue all permit holders with a copy or draw their attention to them.

Water keepers

If a club finds it necessary to appoint keepers to patrol and protect its waters it needs to provide them with an identity document authorizing them to act on the club's behalf and to issue them with strict instructions as to their rights, duties, responsibilities and liabilities. This applies whether the persons appointed are paid or volunteers. A water keeper who does not have this information can be a liability to himself and his club. Provided a keeper has followed his instructions it is up to the club to back him in any action he takes to protect the club's interests, against any angler or poacher; not to do so would be a waste of that person's time and, probably, club money.

The powers of keepers can be strengthened if the NRA appoints them as honorary bailiffs; they would then be entitled to use the powers granted to water bailiffs under the Salmon and Freshwater Fisheries Act 1975. These powers can only be exercised in respect of offences under the Act and cannot be used for other purposes, e.g. in dealing with a person who does not have permission to fish.

Fishery acquisition

A fishery can be acquired either by leasing, under licence, or through purchase and in each case, once the transaction is completed and subject to any conditions imposed as a result, the way in which it is run is the sole responsibility of the tenant or owner.

Occasionally a local authority or business may place a fishery it owns at the disposal of a local club or syndicate to be managed by it on behalf of the owners. Under this arrangement, although the body now responsible may have a certain degree of autonomy, the owners can impose their own conditions on the arrangement within which the club is expected to work.

Value assessment

There are many factors which affect the value of a fishery. Some are obvious (see Fig. 3.4); others are less so and may include such matters as disputed ownership, co-existent rights and planning consents. To ensure that all these are taken into consideration a potential tenant or purchaser is strongly advised to seek professional advice before completing any deal to acquire a fishery.

In making a valuation it is important to obtain as much information as possible about the fishing itself – this can include such things as the catch figures for previous years, potential demand for permits, details of surveys carried out by the NRA, the value of adjoining fisheries. It should not be overlooked that the value could be reflected in the fees that anglers will be asked to pay; agreement may not be reached if these are too high. (The rental value of a fishery usually varies from one twelfth to one twentieth of the purchase value.)

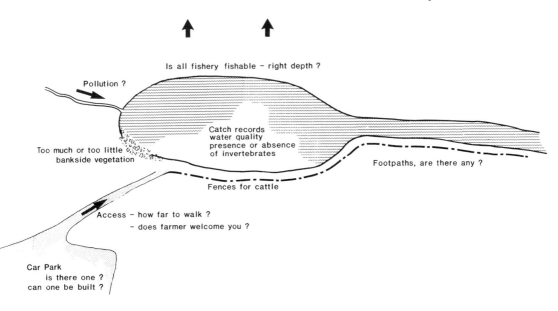

Fig. 3.4 Some of the main factors affecting the value of a fishery.

Leasing a fishery

Although an informal arrangement between a tenant and an owner may seem an easy and cheap way of acquiring the right to fish a water, if there is nothing in writing relating to the terms of the arrangement the tenant has little protection and could lose the fishing at very short notice.

The maxim should always be 'get it in writing' and the best form of agreement is a lease. To try and draw up such an agreement without involving a solicitor or other expert in the field could give rise to a lot of problems.

Prior to the lease being drawn up the potential tenant needs to engage a solicitor and inform him of the various items to be included. It should not be overlooked that the owner will also want conditions put into the lease which will protect his interests.

Some of the items that the potential tenant should consider for inclusion are: protection of tenure on demise of the owner or on change of ownership; the duration of the initial period of the lease and annual rental; an option to renew the lease; conditions for premature termination; determination of the fishery boundaries; rights of access to the fishery and along the banks; the right to catch and remove fish and do work on the bank and in the river; the issuing of permits to the public; stocking and removal of fish; and the control of predators.

The fishery owner will in all probability wish to have all or some of the following covered in the lease: an indemnity against claims arising from the activities of the tenant and requirement that the tenant insures against such;

the date when the rental is due; the right to terminate for nonpayment of rental or other infringement; protection for stock and standing crops; and the right to check credentials of anglers.

Many of these are standard practice but the solicitors will advise if any difficulties arise.

Licensing of a fishery

This is a written agreement between the fishery owner and the tenant which allows him to fish on the water and lays down certain conditions that he must observe. Licensing is a much more satisfactory arrangement than a verbal agreement but provides fewer safeguards than a lease.

Purchasing a fishery

Fisheries can be purchased either at auction or by private arrangement with the seller. The latter is usually the better arrangement although it is sometimes possible to get a very good bargain at an auction.

In view of the many potential pitfalls and legal requirements that relate to the sale of property, and to ensure that once the sale has taken place no unforeseen difficulties or problems threaten enjoyment of the fishery, the buyer's solicitor must be involved in all negotiations.

The purchase of a fishery does not necessarily include the land through which the fishery runs or on which it is located. The seller, therefore, will wish to make certain provisions in order to protect his interests in the land in the same way as he would if leasing the fishery: thus many items which might be considered when leasing might also be included in a 'conveyance' transferring ownership from the seller to the buyer.

In the future the owner may find that the fishery suffers loss or is damaged due to the action of other persons. To substantiate any claim against those responsible it is necessary to show the changes which have resulted and any fishery owner is advised to anticipate any such eventuality by keeping a 'fishery log' and making a photographic record of the water and of any changes that subsequently take place. This, together with fishing catch records, can form a good base on which to make a claim for damage. Incidentally, once any damage is detected it must be drawn to the attention of the person or body suspected of causing it and this should be confirmed in writing. It may also be wise to bring the attention of the NRA to the position.

Managed fisheries

When a fishery is supervised or run on behalf of the owner it often works within constraints that make effective management difficult.

A fishery in a public park, for example, may well have other activities on it, such as boating, that are incompatible with angling and over which the person or body responsible for the running of the fishery has no control.

Private water supply reservoirs on which some works' clubs control the fishing can be drained, resulting in a loss of fish, or have the water level lowered to such an extent that fishing is greatly diminished or impossible.

Anyone offered, or applying to manage, such a fishery must make quite certain that he knows at the outset what he can or cannot do and the outside activities that are likely to be encountered, and consider all the pros and cons before committing himself.

The big advantage of these fisheries is that the rent asked is frequently very small or even non-existent!

Financial assessment of day ticket waters

It is important for most tenants or owners of fisheries to at least meet the annual running costs from the income generated and to adjust the costs of fees or permits accordingly. This must be of prime concern before the lease or purchase of a fishery is undertaken as its viability can determine whether the financial burden that will be incurred merits the expenditure.

Still-water trout fisheries

On a still-water trout fishery up to 80% of total expenditure can be accounted for through the purchase of fish of takeable size. The balance is made up of items such as keepers' wages, rates, property and ground maintenance, advertising, depreciation, and loan and bank charges. The cost of all these, and any others which might occur, represents the minimum expenditure that is required to run the fishery. This must then be related to the income that needs to be generated to meet it.

As an example, a fishery with a bank length of 1200 m and an annual expenditure of £x will need to raise at least this amount, primarily through the sale of permits. Allowing 60 m of bank per angler the fishery can accommodate 20 anglers at any one time but this number is unrealistic and a more reasonable figure would be to assume a 50% uptake of ten visits per day throughout the year, i.e. a total of 3650. If this is the basis on which permits are to be charged each angler would be required to pay £x ÷ 3650 per visit. This figure, however, does not take into account other matters which could influence the cost: e.g. if a close season is imposed the number of fishing days is curtailed; if there is boat fishing the income could be enhanced; disease may require the fishery to close for a period. All of these, together with an emergency reserve, should be allowed for in assessing the income potential.

Still-water coarse fisheries

The maximum number of anglers that can be accommodated on a coarse fishing lake is dictated by the length of bank which is directly related to its shape (a 1 ha circular pond will have a bank length of 350 m, but the same area in the form of a rectangular pond, 50 m by 200 m, will have a 500 m bank length). It is necessary to allow a minimum of about 15 m between anglers which, if divided into the total bank length, would give a figure for the maximum number that could be accommodated at any one time. Again, to be realistic, an uptake of less than 50% should be allowed for.

Most coarse fisheries, once stocked, do not require an annual stocking and this means that the running costs are far less than those of a trout fishery of the same size; apart from this, however, all the other items of expenditure considered for trout fisheries have to be taken into account when estimating the income potential.

River fisheries

The numbers of trout fishermen that can comfortably fish a river, which affects the potential income, is restricted by the nature of, and access to, the water. High banks, heavy vegetation, trees and bushes, excessive weed growth and long walking distances are all major discouragements to an angler, especially if he wants to fly-fish.

A good coarse river fishery may command the same price as a trout fishery of comparable size, but its value is influenced by the supply and demand for fisheries of this type. Such a fishery near a conurbation would obviously have a higher value than that of a similar fishery far from cities or towns.

3.3 Commercial exploitation of coarse fish

Farming carp for food was a widespread practice until, at the end of the nineteenth century, marine fish became freely available in places distant from the sea. In recent years there has been a small but constant demand for carp, eels, bream, roach and a few other species for food, particularly in large towns with a European immigrant population.

Only a few of the anglers using rod and line retain coarse fish for food. Any of the methods described in Section 1.6 could be used to capture and remove them but because of low demand very few are employed. Several specialist types of gear have, however, evolved to catch elvers and eels.

Elvers

The European eel spawns in the Sargasso Sea in the southwest Atlantic. The larval eel (leptocephalus) is carried by water currents for 3 years, and then changes into an elver as it approaches the continental shelf. (It should be noted that some recent reports may indicate a much shorter migration period than was previously thought.) It then migrates up the western coast of Europe from Portugal and is found in greatest numbers in the rivers flowing into the Bay of Biscay. Some elvers also enter the Mediterranean and are caught in Italy and Egypt. The seasons and size of elvers varies considerably (see Table 3.1).

Table 3.1 Seasonal size of elvers in different regions

Country	Time of Year	Number of elvers per kg
Portugal and Spain	December/January	2700
French Coast	February	2800
Britain, Denmark, Holland	April/May	3500
Egypt	January/February/March	4500

Most British rivers have an elver run, and the best known of these is on the River Severn. As the elvers enter the estuary they are transparent, and at this stage are called 'glass eels'. Here they undergo a number of behaviour changes and show maximum activity during darkness. They change again as they pass up river and become progressively more pigmented ('black elvers'). Active migration does not usually occur below 6–8°C. The largest surface migrations occur in conjunction with phases of the moon, thus coinciding with spring tides and producing a characteristic rhythmic 14-day pattern. Vast numbers of elvers enter the Severn and other rivers in each spring tide cycle between March and mid-May, and with each tide of the cycle the elvers move progressively upstream. As the flood tide passes up the river, the elvers appear to be randomly distributed, but as soon as the tide ebbs, the elvers respond to the flow of the river and become concentrated in a ribbon against each bank. It is at this time that they become vulnerable to fishermen using the traditional hand-held elver net (Fig. 3.5). This is similar to a scoop, and is made of cheesecloth or similar material stretched tightly over a willow frame approximately 1 m long, 0.5 m wide and 0.5 m deep. The net is dipped into the river and held there for a few minutes before being lifted and drained; the elvers being tipped into a bucket. They are then placed in flat muslin-based stacking trays, each holding 2 kg, before being taken by the fishermen to one of the elver stations for sale, storage and distribution. Any debris and dead elvers are removed, and the live elvers are stored in recirculating systems (although prolonged storage leads to weight loss).

Elvers caught in recent years have gone principally to the European mainland for restocking. The market in the British Isles is also for elvers for restocking with a big decline in fish for growing on in eel farms. Annual catches vary but the range is 5–60 t for the Severn. Other elver fisheries (rivers Bann, Wye, Usk, Parratt) together account for another 5–6 t per annum.

Since the early 1980s, there has been an apparent decline in elver catches and stocks both in Britain and on the Continent. Many reasons have been advanced for this, but there is no clear conclusion as to the cause. Active conservation measures are being practised including river restocking and the construction of elver passes on weirs.

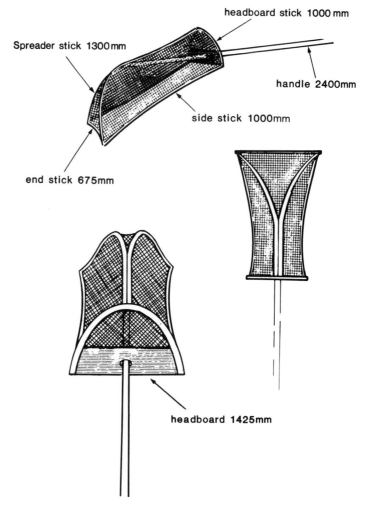

Fig. 3.5 The River Severn elver net.

Eels

The adult eel lives in a wide range of habitats throughout the British Isles, from upland trout rivers to estuaries and from drainage ditches to large eutrophic lakes. Most seaward-migrating eels (silver eels) are 9–12 years of age, having spent 3 years migrating as leptocephali and 7–10 years residing in fresh water during which time they are known as brown or yellow eels.

Before migrating, yellow eels undergo several body changes that equip them to survive in the sea. Once of these involves laying down a deposit of silver pigment (guanine) below the skin surface. From August to November, silver eels descend towards the sea, and large migrations tend to occur on dark stormy nights during which time they become vulnerable to trapping. The best locations are at the outlet of a lake, or at the bottom of a river system, with highest catches in tidal reaches. The fixed weir trap and wing net exploit the silver eel run, while putcheons are used to take yellow eels: fyke nets may be set to take both yellow and silver eels.

Wing nets

The wing net (Fig. 3.6) has a pair of leaders to guide the eels into the net. The net itself is in the form of a long tapering sleeve with two or three inner 'non-return' valves, supported by hoops and leading to a detachable cod-end. The mouth of the net is square or circular and is kept in shape by a frame or hoop. Size varies considerably from locality to locality, but in the main River Severn fishery the nets are extremely large. They are used in the area around Gloucester where the river is about 80 m wide and are set from bank to bank in order to fish the main channel. Two men and a boat are required to operate them.

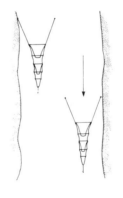

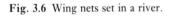

Fig. 3.6 Wing nets set in a river.

Dutch fyke nets

Fyke nets (Fig. 3.7) consist of two parts, the net proper and the wing or leader. The net is conical, the opening (circular or D-shaped) varying in size from 0.25 to 1.2 m in diameter and up to 4 m in length. The smaller sizes are more commonly used in stillwaters for yellow eels while the larger ones are used in tidal waters for silvers. Inside the conical net are funnel traps leading to the cod-end, each trap being progressively more constrictive in aperture. The wing or leader consists of a rectangular wall of netting of about 12 mm knot to knot, fastened to a head and foot rope. Wings, leaders and cod-end are staked. All fyke nets should be licensed and tagged and fitted with an otter guard.

Fyke nets are commonly set in gangs (Fig. 3.8) with a pattern of leaders set out to intercept eels moving both upstream and downstream and, in the case of a lake, along the margins. Net openings may be set facing upstream, down-

Fig. 3.7 Fyke net.

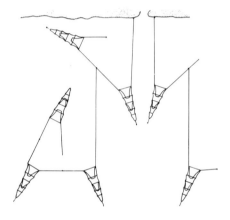

Fig. 3.8 Fyke nets set in gangs.

stream or at right angles to the shoreline. In tidal waters, nets are set so as to be exposed at low tide. In non-tidal waters, cod-ends have to be lifted in order to remove the catch.

Putcheons

Baited eel traps, or 'putcheons' formerly constructed of withy (*Salix* spp) are now made of wire and are fairly extensively used to catch yellow and some silver eels between June and September. High river flows are preferred, and the traps may be set for about 90% of the 4 month period, being lifted every 2 to 5 days. Inside each trap are fitted two constricted throats through which the eel must pass to reach the bait chamber. In non-tidal reaches the putcheons are usually laid with the trap mouth upstream to catch silver eels, but in tidal reaches they can be laid facing either direction to catch eels moving up and down with the tide. Baits used are commonly animal offal, such as rabbit, or fish such as gudgeon and lamprey.

Weir traps

Many traps are now located in disused mill races rather than in weirs con-

TYPE A

THE MORE MODERN TYPE OF EEL TRAP WHICH SUPERCEDED THE OLD
TYPE OF **GRID** SYSTEM. THIS TYPE OF TRAP IS USUALLY INSTALLED
WHEN MILLS CEASE TO FUNCTION AS MILLS, AND THE TURBINES OR
WHEELS ARE REMOVED. THE TRAPS ARE EASY TO INSTALL AND MAINTAIN

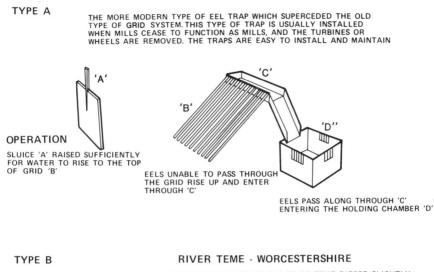

OPERATION

SLUICE 'A' RAISED SUFFICIENTLY
FOR WATER TO RISE TO THE TOP
OF GRID 'B'

EELS UNABLE TO PASS THROUGH
THE GRID RISE UP AND ENTER
THROUGH 'C'

EELS PASS ALONG THROUGH 'C'
ENTERING THE HOLDING CHAMBER 'D'

TYPE B   **RIVER TEME - WORCESTERSHIRE**

ALTHOUGH THE DESIGN OF THE TRAPS ON THE RIVER TEME DIFFER SLIGHTLY
TO THOSE OF THE RIVER AVON, THEY WERE EQUALLY EFFECTIVE, THEY DID
HOWEVER, REQUIRE MORE ATTENTION WHEN 'FISHING' BEING USUALLY SMALLER AND,
SUSCEPTIBLE TO BLOCKING BY DEBRIS DURING SPATE AS NO HOLDING FACILITIES
WERE INCORPORATED, SEPARATE CAGES WOULD HAVE BEEN NECESSARY.

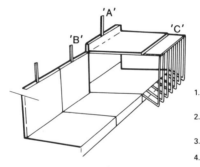

OPERATION

1. SLUICE 'A' OPENED TO ALLOW
   WATER INTO TRAP

2. EELS UNABLE TO PASS THROUGH
   GRID, REMAIN WITHIN TRAP

3. SLUICE 'B' OPENED, 'A' IS CLOSED

4. CATCH REMOVED BY MAN,
   AT APERTURE 'C'

TYPE C   VARIATIONS OF TRAPPING AND HOLDING CHAMBERS,
                           COMMONLY USED ON BOTH RIVER AND LAKE TRAPS

Grid, Movable or Fixture

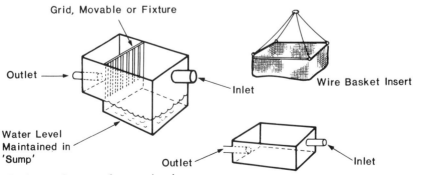

Outlet

Inlet

Wire Basket Insert

Water Level
Maintained in
'Sump'

Outlet   Inlet

Fig. 3.9 The three main types of automatic eel traps.

structed primarily to take eels. The essential feature of all these automatic weir traps is that descending silver eels are intercepted by some form of grating and are deflected into an adjacent holding facility for collection (see Fig. 3.9). Three basic types are used.

Type A: The most common trap used in abandoned mills, where it is fixed in place of the mill wheel or turbine. A sluice gate controls the level of water in the trap channel so that the upstream edge of an inclined grating is just awash. Any eels moving downstream cannot pass through the grid, but are forced to wriggle over the top and into a trough which eventually leads to a holding chamber. It is important that the spacing of the grid bars should not exceed 14 mm.

Type B: Although the principle and design of these traps is only a little different from the previous type, they are much more prone to blocking with debris. They are constructed in ranks, with dividing walls between each.

Type C: This type is more commonly used at lake outlets, although it could equally well be used in a disused mill. The inlet pipe feeds directly into a holding chamber or wire basket, preferably through a grid system to prevent clogging with debris. The outlet pipe controls the water level in the holding chamber.

Other traditional methods

Several other eel-catching methods have been used in the past, and some are still used today. These include 'patting' or 'bobbing' (using worms threaded onto worsted twine); 'sniggling' (using the principle of the gorge bait); baited night-lines; the putt (a large basketwork salmon trap or fixed engine, see Fig. 3.15), operated in the lower Severn estuary which can be modified to take eels, flounder, mullet and shrimp; and the 'tump-net' which is a small-mesh hand net used for scooping eels from beneath overhanging banks.

Eel fishing yields and marketing

The occurrence of differing eel densities under natural conditions is primarily a function of the immigration of young fish. Superimposed on this are such factors as food availability, predation, space and cover, which govern mortality rates and ultimately production in terms of kilogrammes per hectare of marketable fish. The following statistics for yields (or total catch) are available (Table 3.2).

In those lakes or rivers where eels are either absent or present in very low numbers, stocking of young 'boot lace' eels (20–50 g) or elvers will lead to the

Table 3.2 Eel yields from lakes and rivers

	Lakes (kg/ha)	Rivers (kg/ha)
High yields	10–40	20–50
Average yields	3–10	5–20
Low yields	3	3

development of an eel fishery. Stocking must be on an annual basis and for a minimum of 10 years in order to achieve eels of a marketable size. Stocking rates of 1 kg/ha of 20 g eels should yield 10 kg/ha; and 2 kg/ha of elvers should yield 10 kg/ha.

The type of eel caught is different at various times of the year. The activity of yellow eels is related closely to temperature, with peak catches occurring at the height of summer. Silver eel capture takes place in the autumn, but peak capture times vary according to distance from the sea. There is also a distinct size differentiation between males and females; the average length of males is 35–41 cm and that of females is 54–61 cm.

There are two distinct morphological types present in eel populations – the broad headed and narrow headed. The latter tend to have a much higher fat content and are more suited to smoking. They consequently command a premium price. Head morphology is thought to be related to diet, the broad-headed eels' food consisting mainly of fish and molluscs whereas the narrow-headed eels feed mainly on insect larvae.

European markets have definite size and quality preferences. Most eels in excess of 120 g are saleable.

3.4 Commercial exploitation of salmonids

The main species of salmonid present in the British Isles – Atlantic salmon, sea trout, brown trout, rainbow trout and American brook trout – are all usually taken for food on recreational fisheries. There are, however, well-established commercial fisheries for Atlantic salmon and sea trout which use different catching methods. Some of these methods are unique to specific areas of the country; others are similar to those described elsewhere in this book (see Section 1.6).

This Section will briefly describe the catching methods used by commercial salmon fisheries found near to or within the home river system of the British Isles.

The fishing instruments used can be divided into fixed and mobile types.

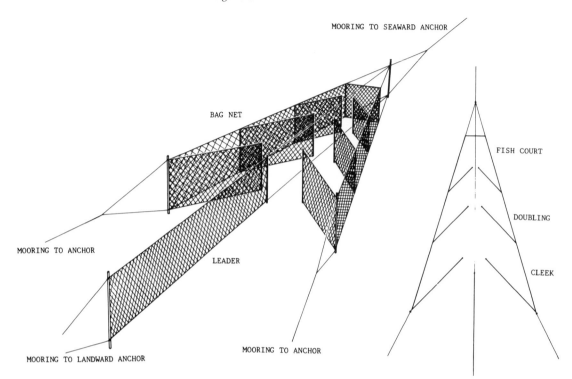

Fig. 3.10 Diagrammatic plan of a bag net, the top and bottom of the net having been omitted for clarity.

Fixed engines

Bag net

The bag net (Fig. 3.10) may be defined as a net extending seawards from the shore, suspended from floats and anchored in a fixed position. It consists of a trap made of netting into which fish are directed by a leader, also of netting. The leader does not usually exceed 120 m in length. One end of it is attached to the trap and the other securely fixed either to the shore or to a stake in the seabed. The material from which the leader is made is of sufficient thickness to be visible to the fish, and assists in directing them. No part of the trap netting should be of thinner material than the leader. Bag nets are often shot in a line extending seawards from the end of a shore-attached bag net or stake net.

Stake net or fly net

The stake net or fly net (Fig. 3.11) is a net fixed to the foreshore by stakes. It may be defined as a curtain of netting erected on stakes and set vertically in the foreshore. It acts as a leader to approaching salmon, with a pocket or trap

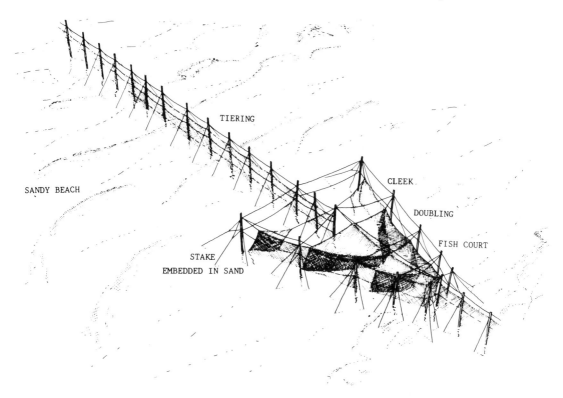

TIERING

CLEEK

DOUBLING

FISH COURT

SANDY BEACH

STAKE
EMBEDDED IN SAND

Fig. 3.11 The stake net, or fly net.

inserted at intervals to take fish which are directed along the leader. It is fixed to the foreshore throughout its length.

Jumper net

The jumper net is a type of fly net in which the stakes and netting of the leader are replaced by a floating curtain of netting which is fixed at both ends and which rises and falls with the tide.

Poke nets

Poke nets (Fig. 3.12) are used exclusively on the Scottish side of the Solway Firth. They are mounted in lines on rows of poles and consist of a series of pockets of net in which fish are trapped and enmeshed.

'T' nets

These nets (Fig. 3.13) are a development of bag nets and operate on the same principle.

Fig. 3.12 Poke nets.

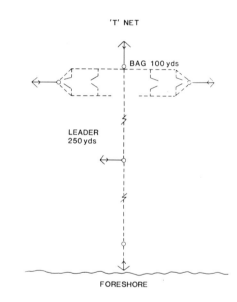

Fig. 3.13 The 'T' net – a development of the bag net – in plan form.

Putchers and putts

The majority of salmon caught in the Severn estuary are taken by fixed engines, the most common of which are putchers (Fig. 3.14). These are typically long, conical basketwork traps (although nowadays virtually all are constructed of plastic-covered steel or even stainless steel) about 1.5 m in length and 0.6 m in diameter at the opening. Several hundred may be mounted on a framework of larch and elm poles to form a 'rank' or fishing weir. Specific sites only can be used now for putchers, principally those certified in pursu-

Fig. 3.14 A rank of putts and putchers on the River Severn.

ance of the 1865 Salmon Fishery Act and known as 'privileged fixed engines', or occasionally those which are lawful by virtue of grant, charter or immemorial usage, providing they were also fishing in the open season of 1861.

Most putcher ranks fish the ebb tide and trap fish descending the river during their period of estuary residence before final upstream migration into freshwaters. Some ranks, however, do fish the flood tide. Efficiency of the fixed engine is greatly affected by tidal height and range.

A putt (Fig. 3.15) is a much larger trap than the putcher and consists of three separate sections – the kype, butt and forewheel. The kype (outer end) may measure 2 m in diameter with the whole trap being about 4 m long. Putt weirs are positioned with the mouth upstream to fish the ebb tide. Salmon, eels, dabs, flounders and shrimp are all caught in putts because the close weave of the forewheel and butt (middle and end sections). Very few are used now.

Stopping boats

None of these boats survive in active use today. (In 1866 the Commissioners for English Fisheries issued Certificates of Privilege for 24 stop nets.) Stopping boats are wide-beamed, stoutly built boats up to 7 m in length, with a shallow draft. They are moored at right-angles to the tide, attached to a cable. A number of boats may be attached to the cable in order to fish the channel, generally on the ebb tide, although there are also flood tide stations. Long poles are lashed in the shape of a 'V' with the apex inboard, and to this frame is attached a large bag net. The whole structure is finely counterbalanced and

Fig. 3.15 A single row of putts.

supported in the fishing position by a prop. A number of cords attached to the bag of the net are held taut by the fisherman so that he can feel when a fish hits the net. The prop is then knocked out, bringing the counterbalance into operation. The arms (raines) lift up, bringing the net opening above the water surface and trapping the fish. The end of the net is then pulled to the surface and the fish is removed and killed.

Mobile fishing instruments

Seine, draft, or net and coble

These synonyms are used in various parts of the British Isles to describe an encircling method of fishing with a net and boat (Fig. 3.16). In Scotland, under the Salmon Fisheries (Scotland) Act 1868, the net and coble is the *only* method of netting permitted within the estuary limits. These limits, however, as shown in Schedule B of the 1868 Act, delineate only the seaward limits of the estuaries, and the inland waters so contained may include seawater lochs, rivers and freshwater lochs. Legal net and coble fishing must comply with certain rules which have been established by precedent in the Scottish courts. These decisions require that the net must not leave the hand of the fisherman and must be kept in motion relative to the water while fishing. The construction of the net must also show that it is designed to encircle the fish and not merely to enmesh them.

Fig. 3.16 The net and coble fishing method.

The term draft or seine is applied to numerous similar fisheries in England and Wales. However, their numbers have declined substantially since the middle of the last century, almost entirely due to byelaw limitations. In the Severn, for example, draft netting extended almost as far as Welshpool on the upper reaches, but progressive curbs have limited their numbers, all located well below Gloucester in the tidal reach. High river flows or spring tidal velocities will often prevent successful netting by interfering with the set of the net.

Haaf nets

The haaf net (Fig. 3.17) is restricted in use to the Solway Firth. The net is mounted on a wooden frame about 5 m × 1.25 m and the fisherman stands in the tide with the 'middle stick' over his shoulder and the net grounded in front of him. When he feels a fish strike the net, the frame is twisted, trapping the fish. Haaf netters frequently fish in groups, in line across the channel, on both flood and ebb tides.

Lave nets

The lave net is very similar in principle to the haaf, except that the net is

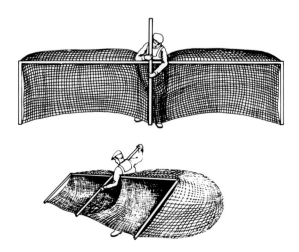

Fig. 3.17 The haaf net, used only on the Solway Firth.

suspended from a collapsible Y-shaped frame about 2.5 m high and with a gap of about 3 m when open. Its use seems to be restricted to the Severn estuary, and there are two modes of operation. The first is similar to that employed with the haaf, but on the Severn this is known as 'laving'; the second can only be operated under certain conditions of wind and sand configuration. It depends on the fisherman seeing the wake or 'loom' of the fish as it heads back from the receding shallow water to the low water channel. The lave netsman then runs towards the fish, sometimes in knee deep water, and scoops it up. Fish caught in this way have been stemming the ebb or moving upstream on the ebb.

Drift nets

Drift nets consist of either a single wall or multiple walls of netting, commonly 2.5 m but sometimes up to 6 m deep. They are shot from a boat across the current and are allowed to drift freely. Nets are up to 370 m long in most inshore and estuarine fisheries, but may be considerably longer in coastal or offshore fisheries. Single-wall nets are simple gill nets with mesh sizes ranging between 60 mm knot-to-knot for grilse, up to 90 mm knot-to-knot for two-, three- or four-winter sea-fish.

Multiple-wall or trammel nets (Fig. 3.18) may consist of two or three walls of netting, one of which (the lint or linnet) is made of small meshing, the other(s) (the armour) of large mesh. The principle of the trammel net is to entangle. The fish swim through the large meshes of the nearer outer wall, carrying the very loosely hung small-mesh inner netting through the larger meshes of the farther outer wall, so trapping themselves in a pocket. These nets are operable on both flood and ebb tides.

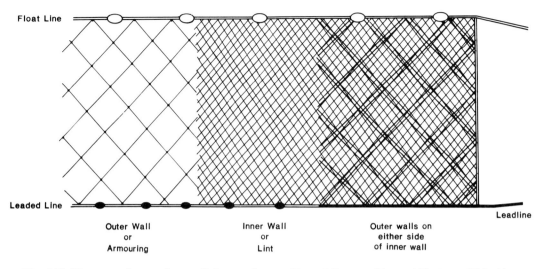

Fig. 3.18 The trammel net, a three-walled entangling net (From J. Burgess, *Trammel Netting*, published by Bridport Gundry Ltd., Dorset)

Other instruments
Several other fishing methods are used for salmon, including coracle and compass nets (Wales); stand, bow or click nets (Humber); snap, pole and hoop nets (Ireland); and channel, yair and shoulder nets (Scotland).

3.5 Aquaculture of salmonids

There are many publications for the fisheries manager to read concerning the history and development of the various aspects of aquaculture. Although it is most unlikely that any 'fish farmer' has learned from a book how to rear fish, it is hoped that the more obvious pitfalls can be avoided by the written word. This section, therefore, attempts to bring to the potential fish farmer's attention some practical advice that may make life a lot easier. (This Section only relates to England and Wales. The legislation is different in Scotland and Northern Ireland.) It is only from practical experience on the site chosen that problems peculiar to that location will be identified.

Aquaculture in the British Isles has been, and still is, principally concerned with the rearing of salmonids. Recently, however, there has also been much development in the rearing of coarse fish such as carp and eels, and other species such as crayfish. There were approximately 650 fish farms in England and Wales in 1991. These produced about 8000 t of rainbow trout for the table and an unrecorded quantity of other species for restocking. Relatively small quantities of carp and eels were produced for the table but in Scotland more than 40 000 t of salmon were produced for human consumption.

Well-established markets exist for trout both for the table and for restocking. For the table, rainbow trout are produced in greater numbers than brown trout, and fish 20–23 cm long and weighing approximately 180 g, or 25–28 cm long and weighing 228 g, are the normal sizes sold. There is also some demand for larger fish up to 2 kg. A wider size range can be sold for restocking and the higher quality demanded commands a higher price. Cage-reared salmon, particularly from Scotland, are now successfully sold in the same market as commercially caught fish, and in far greater quantities.

Salmonids can be reared (a) on land-based sites or (b) in floating cages. Rearing and husbandry techniques have evolved to suit the different requirements demanded in the life-cycles of rainbow trout, brown trout and salmon. Fig. 3.19 shows that the three species, at least in freshwater, grow at different rates. The following descriptions relate principally to the factors that should be taken into consideration when choosing a site, and also to the day-to-day routines for trout, with examples of the requirements for rearing salmon to the smolt stage where these are different.

Salmonid aquaculture on land-based sites

Water supply

There are two main types of water supply suited to the production of trout and young salmon: underground (or borehole) water and surface water. Borehole water generally has a temperature of 9–10°C and often contains very little dissolved oxygen. It has advantages for egg rearing in that the temperature is higher during the winter months than that normally found in rivers, thus allowing eggs to develop more quickly. It rarely contains high levels of suspended solids, so the eggs are not smothered.

The disadvantages of boreholes are that they are costly to drill and there is no guarantee that a licence will be issued for their use. Borehole water normally needs to be pumped to the surface, and the electricity needed is expensive. The consequences of a power cut may be disastrous, and a second pump and a standby generator are therefore required as a safeguard. Spring water combines all the virtues of borehole water without some of the drawbacks.

If surface water is the main source of supply, the volume of water in the river or stream during low flows, such as prevail during a severe drought, will determine the output of fish from the farm. It is important, on a site fed by surface water, that there is sufficient drop in the level of the land to allow a gravity feed onto the site and also a gravity outfall for drainage on the downstream side. If the water has to be pumped, this can be very costly. Land-based salmon farms can utilize the sea water pumped into tanks to emulate the marine growth phase.

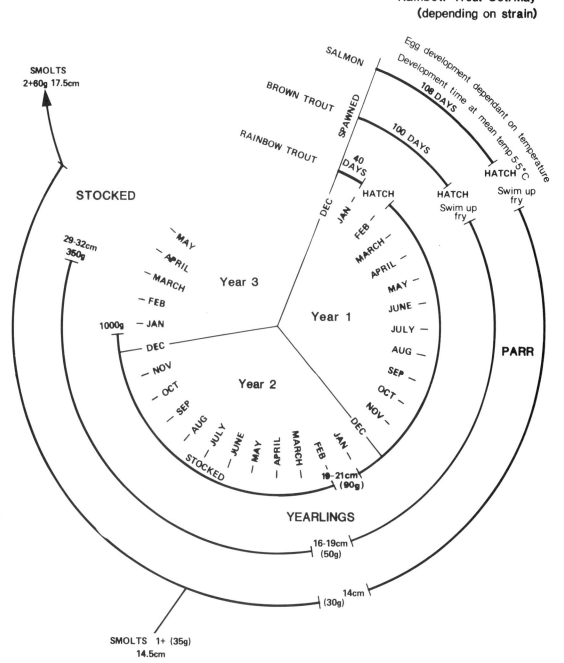

SPAWNING TIMES

Salmon December
Brown Trout Nov/Dec
Rainbow Trout Oct/May
(depending on **strain**)

Fig. 3.19 The rearing cycles of the principal salmonids.

Table 3.3 Optimum water quality requirements for trout farming

Biochemical oxygen demand	1.3
Suspended solids (mg/l)	6 (increasing after rainfall)
Total oxidized nitrogen (NO_3)	2.7
Ammonia (mg/l as N)	0.1
pH	7.3 (min 7.1–max 8.3)
Temperature °C	4–18

Variations to these values may be acceptable depending on the size and method of culture chosen.

Water quality

The water used for trout farming needs to be of constant high quality (Table 3.3).

When choosing a site, great care should be taken to ensure that if a surface source is to be used, it is at minimal risk from pollution. Streams that receive effluents from sewage works or trade premises (including quarrying and gravel excavation), or drainage from main roads, should normally be avoided. The water supply should be well aerated, and have very little BOD or free and saline ammonia. Borehole supplies should be checked for supersaturation with nitrogen gas, which would require dispersion.

The site and locality

To accommodate the necessary ponds, access roads, accommodation and other buildings (Fig. 3.20) the minimum land area required is one hectare for every 10 tonnes of fish-production capacity. However, this requirement depends on the type of unit employed. The site should be free from flooding, and have an adequate fall so that all ponds can be drained individually. The security of the site against vandalism and theft is another point for consideration.

It is important to take into account the proximity of the market to potential sites, and the quality of the road system. Special equipment to transport live fish from the farm to the fisheries is essential (especially if long distances are involved) since safe transport of stock is the responsibility of the vendor. There are, however, firms specializing in the transport of live fish.

Manpower

A minimum of two full-time staff is required for a 20 t unit. It is very useful to have someone living on or near the fish farm to cope with emergency failures in pumps or aerators, sudden pollution, flooding, or accidental pond drainage. A person living on site also gives added security against theft or vandalism.

Culture and husbandry techniques

Some trout farmers breed from their own stock, but this is not always

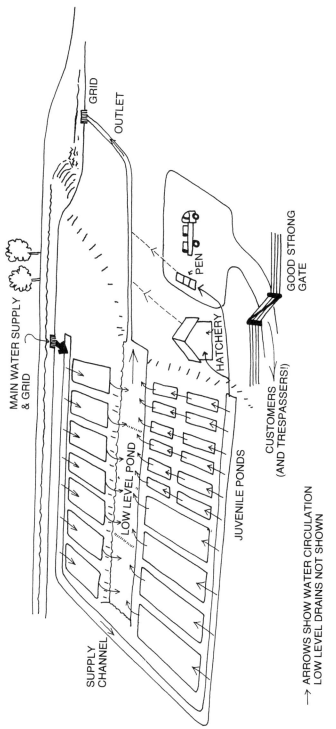

TROUT FARM-EARTH POND VERSION

GRID

OUTLET

MAIN WATER SUPPLY
& GRID

PEN

GOOD STRONG
GATE

HATCHERY

SUPPLY
CHANNEL

LOW LEVEL POND

JUVENILE PONDS

CUSTOMERS
(AND TRESPASSERS!)

→ ARROWS SHOW WATER CIRCULATION
LOW LEVEL DRAINS NOT SHOWN

Fig. 3.20 Trout farm – earth pond version.

worthwhile for the small producer. An increasing trend is for eyed eggs (ova) or fry to be brought in from specialized suppliers. Farmers should consider the use of genetically manipulated stock to avoid inefficiency in growth resulting from energy wastage in the development of unwanted sexual products (see Appendix 9).

Stripping

It is advisable to strip male and female fish under cover and in dry conditions, and this should be performed out of direct sunlight to prevent ultra-violet rays from the sun reducing egg survival rates. Some farmers strip eggs into muslin before tipping them into a dry bowl for fertilization with milt. This can prevent ovarian fluid from an overripe trout coagulating the milt and so preventing fertilization. A 1 kg trout will produce 1500–2000 eggs, and usually the eggs of three female trout are mixed with the milt of one male. For salmon the sex ratio may be 2:1 female to male or 1:1, since a large female may produce many thousands of eggs and take some time to strip. Eggs and milt are then mixed with the fingers and clean water is added to aid fertilization. Excess milt is carefully washed away, and the fertilized eggs are left to water-harden for up to 1 hour, after which they are counted out into trays.

From this stage onwards the eggs become increasingly sensitive, and after 48 hours they should not be moved. At a temperature of 10°C rainbow trout eggs will hatch in about 30+ days. Salmon development is slower. December eggs will not hatch until March in the south of England, and later in Scotland. Approximately halfway through their development the embryo will develop eyes, and once these are clearly visible the eggs can be handled again. At this stage the eggs are 'shocked' by shaking to sort out any unfertilized ones which become opaque and can thus be picked out. Eggs require about 20–25 litres of water per minute through each trough. Ideally it should be of good quality, free of suspended solids and with a constant temperature.

Alevins

The alevins hatch out and either drop through the egg trays or are retained, depending on the type of tray in use. There can be severe losses at this stage due to suffocation, as the alevins tend to seek cover and huddle together in corners. The baskets used for the retention of alevins are designed to provide more corners for them, thus helping to prevent suffocation.

Alevins use up their yolk sacs after a few weeks (4 weeks at 10°C) and then begin rising from the bottom in search of food, hence the term 'swim up' stage. This is a very important period when they must be fed, and food should be offered every 20 minutes throughout the day to ensure that it is available whenever they swim up. Salmon fry begin feeding at a lower temperature than trout (about 7°C) and this can create some problems of overcrowding at indoor sites fed with borehole water and with outdoor growing-on facilities. Low outdoor temperatures may prevent an early transfer of fish from indoor tanks until later in the year. There must therefore be sufficient indoor holding capacity to cope with the fry populations until transfer to outside tanks is possible.

Fingerlings

Young fry can be grown in hatching troughs for about a further 3 weeks and are then usually transferred to circular or square tanks. They are stocked at high density – 20 000 fingerlings to a 2-metre diameter tank with a water depth of 30 cm. In the early stages this high density actually improves their feeding and growth. The water supply to the tanks should be about 120 1/min. The fingerlings are thinned out as they grow, and the water depth is increased.

Fish are stocked out from the hatchery at 12 cm-plus in length. By this time, whirling disease, if present in dirt ponds, will not affect them by causing brain damage and premature death as their crania are sufficiently hard to prevent damage. Salmon parr would be 5–8 cm in length by mid-summer.

Salmon

The fastest growing fish of 12.5–15.0 cm will become smolts at 1+ years of age in April or May of the year following at an average weight of 35 g. The smolt stage is characterized by a silvery appearance to the sides of the fish caused by the deposition of guanine on the scales. Other physiological changes occur to the fish which enable it to adapt from freshwater to seawater. Approximately 40% of salmon stock will smolt at 1+ years old. Nearly all the second-year fish should become smolts at 2+ years old. The average weight of a 2+ smolt will be 60 g. Survival rate for salmon from egg to smolt is likely to be only in the order of 50–55% in ideal hatchery conditions.

Stocking densities and water supply

Ponds

Earth ponds can usually hold from 8 to 30 kilogrammes of trout per cubic metre, circular tanks from 30 to 40 kg/m^3, and raceways also from 30 to 40 kg/m^3. The water supply has three main roles: (1) to supply oxygen, (2) to take away waste products, and (3) to support the fish. The quantity of water required per pond will vary with temperature and weight of fish present. Ten megalitres per day is the minimum amount required to produce 25 t of trout in a one-pass system. A unit producing 25 000 smolts a year will require a maximum water flow of 4.0–5.5 megalitres/day in a one-pass system.

Feeding

Trout pellets are readily available and 1.5 t of food is required to produce a tonne of fish. (The actual conversion efficiency of food protein to fish protein is about 4.5 to 1, because the food has three times the concentration of protein found in fish flesh.)

For rainbow trout, the average production cycle is about 1 year (Fig. 3.19). This consists of up to 2 months for hatching, 1 month in the alevin stage, followed by 9 to 10 months growing to marketable size. This period may be longer in colder waters.

Grading

The individuals in a population of fish grow at vastly different rates, and grading is necessary whenever there is a marked size variation among fish grown together. Grading, however, should not be undertaken too often, since growth will be retarded and fish may be damaged. Many farmers grade only twice a year. Grading breaks up hierarchies established by dominant fish and improves feeding efficiency and growth rates because fish get the correct pellet size. Continuous market production can be achieved throughout the year because of this difference in growth rate.

Financial aspects

Anyone intending to establish a new fish farm or purchase an established unit should follow the basic financial rules for establishing any business. There are many professionals who can advise on raising the finance, and it may be worth having a consultant rather than risking financial losses.

The following observations are presented to illustrate the percentage spending on different items on a fish farm producing 25 t of trout.

*Typical expenditure items**	%
105 000 6–8 cm fish	6
Food 37.5 t	31
Labour, one man full-time	13
Vehicle	5
Insurance	5
Medication and miscellaneous	5
Marketing, advertising, etc	4
Financing charges at 15% interest and repay on initial capital over 10 years	27
Interest on working capital	4

* It has been estimated by Landless (personal communication) that after a site has been purchased £2000 (1982 baseline) of setting up capital is required for every tonne of fish produced, i.e. allowing for construction costs, pipework, building works, etc. A 25 tonne unit would consequently cost £50 000 with an additional £15 000 working capital.

Income

The essence of profitable trout farming is marketing. The selling price at the farm gate or to hotels/restaurants can be 25% and 50% more than at wholesale. Selling all stock at the wrong price to the wrong market can result in a substantial loss. The time involved in the actual handling of sales must also be taken into account.

Legal aspects and other regulations

There are certain legal requirements and regulations governing the setting up of fish farms. Potential fish farmers in England and Wales need to contact their local NRA Region in relation to the following matters.

Abstraction licence

An abstraction licence is required if the fish farm utilizes either borehole water or surface water and exceeds 20 m^3 of water in any 24 hour period. An abstraction licence can be obtained from the NRA's licensing section, who also advise on licensing matters. A new licence, or variation of an existing licence for other purposes, will require advertisement to safeguard the interests of other water users. An annual charge is payable to the NRA for the abstraction rights conferred by the licence. (The granting of an abstraction licence does not imply that the water quality is suitable for fish farming.)

Impounding licence

An impounding licence must be obtained before the construction of any weir or dam takes place in a watercourse for the purpose of diverting the flow into the fish farm. The licence is required in addition to any licence for abstraction, and application is again made to the NRA. In certain cases where water is to be impounded, the Authority will also require a special agreement to ensure adequate protection of downstream interests. A large impoundment of more than 25 000 m^3 is subject to additional control to ensure the safety of the dam, and specialist engineering advice must be obtained from a Panel 1 Engineer appointed under the Reservoir Safety Act 1975.

Land drainage consents

NRA Regions have established land drainage byelaws which require consent for certain operations in or adjacent to a main river, or any watercourse flowing directly thereto, and in the river flood plain. (The use of the term 'main river' in this context relates to the provisions of The Land Drainage Act in which certain rivers may be designated as 'main river'. It must be emphasized that quite small rivers are sometimes 'mained', and the advice of the NRA's Flood Defence departments should be sought.) Such operations include erection of fences, tree planting, disposal of rubbish, excavation affecting the bed or banks of rivers, erection of jetties and walls, etc. The Authority's consent is also required for the erection or alteration of any structure in, over or under a main river which might be likely to affect the flow or impede land drainage works.

In the case of 'non-main river' watercourses, consent is required for the erection or alteration of any mill dam, weir, culvert or other obstruction to flow.

In addition to land drainage consent, consent may also be required for certain operations by Internal Drainage Boards and Local Authorities. It is important to remember that in order to dispose of waste water, there should be a stream or watercourse on, or adjacent to, the site where a fish farm is to be constructed.

Consent to discharge effluent

The effluent and any discharge from a fish farm falls within the definition of trade effluent. Discharge of effluent to a watercourse therefore requires the consent of the NRA Region. The quality conditions imposed on each individual discharge will be determined by the particular requirements of the receiving stream, but the following guidelines should be used:

- If possible, there should be a single outlet for the discharge, and facilities must always be provided to enable samples of the effluent to be obtained.
- Provisions may be required for measurement of the volume of water abstracted and discharged.
- The quality of the effluent discharged will be required to be substantially of the same quality as the abstracted water.
- Nothing should be added during the fish farming activities to cause the discharged effluent to be toxic to fish, fish spawn, the food of fish, or to any other river life.
- Prior to ponds being emptied into a river or stream or sterilized, the NRA should be informed.

Under normal circumstances, all these requirements can be met by good husbandry. It should be pointed out, however, that drainage of ponds which have been in use may require settlement to reduce the suspended solids before being discharged to a stream.

The NRA has introduced a charging scheme to recover the cost of operations for control and monitoring of consented discharges and their receiving waters. An application charge and an annual charge are payable. These charges vary according to the volume and quality of the effluent and the type of water the effluent is discharged to. A full discussion of details with the local NRA office is advised for any new entrant to fish farming.

Control on fish movement

The Salmon and Freshwater Fisheries Act 1975 was amended by the Salmon

Act 1986 to relieve most fish farms, except cage farms, from the requirement to obtain NRA consent for the introduction of fish to their premises, provided that the fish farm is registered with MAFF as a fish farm. However, consent will still be needed for introduction to other inland waters and the appropriate application forms should be obtained from the local NRA office.

European Community (EC) regulations (in force at the beginning of January 1993) amended the rules governing the transfer of live fish and eggs between member states and details should be sought from the relevant government department. Live salmonids may not be imported from outside Great Britain whilst salmonid eggs or coarse fish imports require health certification and a licence from the government department (MAFF in England, Welsh Office in Wales).

The Diseases of Fish Acts of 1937 and 1983 lay down systems for the control of movements of fish from the sites infected with certain notifiable diseases. These controls are administered by MAFF and the Welsh Office who will also carry out inspections of disease status, in due course bi-annually.

Planning consents

Prospective fish farmers must obtain permission from the local Planning Authority before carrying out any constructional work. Enquiries should be made from the local valuation officer on the liability to pay rates on the buildings and other works. Guidance on these matters can again be obtained from MAFF. Generally, fish farms producing for food are classed as 'agriculture', whereas those producing solely for restocking fishing waters are not. In addition fish farms are required to register with MAFF or the Welsh Office who should be contacted for details and application forms. Annual returns of stock produced will be required.

Salmonid aquaculture in floating cages

Rearing fish in cages is a well known method in some parts of the world, particularly the Far East, but only in recent years has the practice developed in Europe, and it began mainly with the culture of salmon in the sea in Norway and Scotland. Now several water companies in England and Wales, and some commercial fish farmers, operate cage rearing units in inland and coastal waters, rearing trout for restocking and trout and salmon for the food market. Although basic husbandry techniques are the same, the method is usually cheaper than at land-based sites because the capital costs are low and no land purchase is involved. Running costs are relieved of the burden of screen cleaning, but feeding may take much longer and grading, cleaning and mending the nets are also time-consuming.

Rainbow trout are ideal fish for this type of culture and can be held at densities of up to 10 kg/m^3 in freshwater to produce acceptable fish for the restocking market. In saltwater or when rearing fish for food, as much as double this density is possible. Brown trout can also be reared, but should be stocked at no more than 6 kg/m^3.

Cage types and construction

The basic principle of a fish cage is that a floating bag net is supported at the surface by some form of flotation system and has a framework to keep it in the open position (Fig. 3.21). Types vary from a simple home-made construction of polyvinyl chloride (PVC) piping formed into a watertight square which supports the net (Fig. 3.22), to a ready-made timber or metal assembly with built-in polystyrene float chambers and a walkway/working platform around the perimeter (Fig. 3.23). Flotation can be achieved by using anything from polystyrene foam blocks or buoys, to oil drums or plastic containers. Different sized mesh nets are required for small fry or large fish depending on rearing requirements. The largest mesh possible should be used as this allows maximum passage of water, and the nets are also much lighter to handle. The net needs to be weighted at the corners to keep it squarely open and to prevent it from 'riding up' in storms or strong currents.

Siting and mooring

Sea cage rearing in Norway and Scotland, and more recently in Ireland, has developed successfully because of the ideal conditions in the sheltered lochs

Fig. 3.21 View of cage rearing unit at Clywedog Reservoir. Note the automatic feeders and storage sheds.

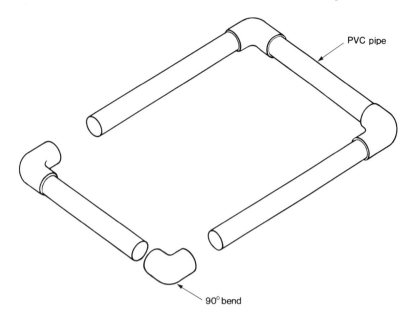

PVC pipe

90° bend

Fig. 3.22 Simple floating cage frame constructed of PVC piping.

and fjords that are used as sites for cage units. Damage from storms is less likely in freshwater but as sheltered a site as possible should be chosen, particularly in larger bodies of water such as reservoirs.

If the cage is to have a rigid anchorage from two or more fixed points, then as a general guide the depth of water needs to be at least twice the depth of the cage net – that is, 6 m of water would be required for a net 3 m deep. Otherwise, problems will occur with a build-up of waste products on the bottom causing low dissolved oxygen levels in summer, and conditions will then be favourable for parasites and diseases.

In larger stillwater bodies where there are perceptible currents, the depth is not so important if the Scottish marine type of anchorage system is used: the whole unit is allowed to swing around one anchor with wind or tide. In this case it is only necessary to ensure that the water is at all times deep enough to prevent the cage snagging the bottom. Anchors can be made of concrete blocks, but such weights tend to drag under extreme strain. A better method is to use a large boat anchor or construct one with flukes that will dig into the bottom. Mooring lines should be of adequate strength, and a length of chain at the anchor end will prevent drag in much the same way as the anchor chain, rather than the anchor itself, holds a ship on station.

At smaller sites it may be easier to moor the cages directly to the shore or to build a service jetty. Specialized fish-carrying boats have also been developed to carry fish from the rearing areas to restocking or collection points.

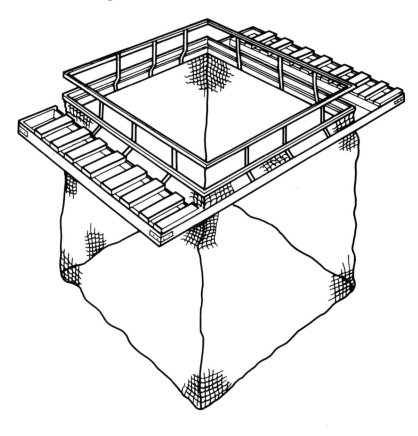

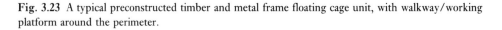

OVERALL DIMENSIONS 6.1m BY 6.1m BY 4.6m DEEP
SIZE OF NET: 5.8m BY 4.6m BY 4.6m DEEP
VOLUME OF WATER ENCLOSED BY NET: 107m^3

Fig. 3.23 A typical preconstructed timber and metal frame floating cage unit, with walkway/working platform around the perimeter.

Water quality and diseases

It has been found that the more oligotrophic waters are best suited to cage rearing. Trout held in cages in lowland reservoirs or lakes with established coarse fish populations tend to become infected with parasites (particularly tape worms and eye flukes) which originate from coarse fish and reach the trout via intermediate hosts.

The water should not be subject to excessively high summer temperatures (20°C or more) for any length of time, and waters with algal blooms should be avoided. In richer, more eutrophic waters, the fouling of the nets by algae can cause problems, and may become so serious that there is no exchange of water through the cage net. This can result in fish deaths from lack of dissolved oxygen.

Routine operations

With experience it will become apparent that certain routine operations are necessary for the successful operation of a cage rearing site. These will depend to a large extent on the size of the unit and where and how it is moored, but will include:

- At least one visit per day when weather conditions allow.
- The feeding of fish twice a day if possible; once in the morning and once in the afternoon, unless automatic feeders are employed.
- Correct feeding schedules. Both underfeeding and overfeeding can cause stress and affect the food conversion (2:1 is normal for cage rearing). The latter also affects the economics of the operation.
- An inspection of the mooring connections, brackets, thimbles and shackles for damage or wear at frequent intervals, particularly after heavy weather.
- The removal of any dead or dying fish for burial or incineration ashore.
- Routine hatchery or fish farm operations such as measuring and weighing samples, grading, etc – carried out in a fashion similar to that for a land-based farm.
- The use of divers for periodic underwater inspections of the unit.

Safety

Working on a floating structure is obviously more hazardous than working on dry land, and personnel should be particularly safety-conscious. Measures to be taken to ensure safe working conditions include:

- Provision of lifebuoys for immediate use.
- The rule that lifejackets should always be worn when afloat.
- Flares or smoke signals, which should be placed on the cages so that if any person becomes marooned, he can signal for assistance.
- Informing a responsible person onshore of each visit (this is particularly important if the cages are in a secluded offshore position).
- The provision of oars on board small powered vessels in case of an engine breakdown.

3.6 Aquaculture of cyprinids

In general, the rearing of coarse fish requires far less water than salmonid farming. They can be grown at higher temperatures in static ponds and they tend to be more disease tolerant and less 'fussy' about water quality. As a result cyprinids can be kept at high densities without aeration and still show good growth rates without supplementary feed being given.

Cyprinid farming began in China as long ago as 475 BC, when the techniques for spawning and growing of carp were first described (Bardach 1972). In Europe, the construction of fish ponds in Czechoslovakia dates back to the eleventh century. This idea originated from the systematic draining and controlling of marsh areas using simple dams which created ponds suitable for fish production. Breeding took place naturally and no attempt was made to control or enhance the stocking of the pond. Gradually a system was developed whereby small numbers of breeding adults were encouraged to spawn in small purpose-built ponds. For the first time control was exercised over the numbers of fry produced and, more importantly, this allowed selection of the traits desirable in farmed table fish. By 1860 carp and other species of coarse fish were being reared for the table in most European countries.

In the last 50 years great advances have been made in the controlled production of cyprinids. Chiefly among these has been the development of hormone-induced spawning or hypophysation. This technique has given the fish farmer the ability to determine when, where and how the brood fish spawn, giving him great control over egg and fry production.

Research into the early feeding of cyprinids and developments in the production of artificial and natural diets has also led to a great improvement in growth and survival rates. A wide range of techniques is applicable to coarse fish farming, from netting a pond every 2 years and cropping what has grown naturally, to induced spawning and growing on in earth ponds or tanks. The choice of system depends upon the level of investment in terms of manpower, money and materials. All the major species of cyprinids have now been reared including chub, dace, barbel, roach, bream, rudd, tench, crucian carp and common or mirror carp. Each of the above can be spawned using variations of the classic carp techniques explained below and most of the major practical problems likely to be encountered will be covered (see Appendix 9 for an annual timetable of cyprinid farming).

Carp

The common carp (*Cyprinus carpio L.*) is a warm-water fish and is the most widespread fish cultivated in the world today. The reasons for this are its ease of culture, fast growth rates, disease resistance and general hardiness with regard to water quality, oxygen levels and temperature fluctuations. The carp is an omnivore, feeding on both vegetable and animal matter, which allows a wide range of natural and artificial foods to be used. Its optimum growth is in the range of 20–28°C which corresponds well with the summer temperatures in the UK. At these temperatures, and with sufficient food, growth is extremely rapid. Temperatures less than 20°C show progressively feeble growth which ceases altogether below 5°C.

Ponds

The size of ponds used in carp culture may vary from 50 m^2 for a spawning or early rearing pond to 5 ha for a growing on pond. In Europe ponds as large as 500 ha may be used although these are an exception. Ponds should be constructed in such a way as to enable them to be drained by gravity. This allows them to be dried and left fallow for a period of time (see Pond Management, page 174). Most are formed by a 'cut and fill' exercise where the pond bottom is scraped out and this material is then used to form the pond sides. Others may be constructed by impounding the pond behind embankments formed from imported materials. Ponds' sides should be constructed with minimum slopes of 1 in 2 and a width across the top of at least 3 m to allow access. The pond bottom gradient must be no shallower than 1 in 70 and this should slope down towards the outlet chamber or 'monk'. These monks can be constructed from timber, brick, glass fibre or more preferably concrete, and should incorporate a screen to prevent the escape of fish. A method of lowering the pond level in a controllable manner using either weir boards or a sluice valve (or better still, both) should also be included.

Water supply

In general, carp farming requires less water than salmonid farming. This is because most cyprinids are grown in static ponds and water is only required for the initial filling of the pond and subsequently to make up any seepage or evaporative losses which may occur. Some fresh water may also be required if the ponds need to be 'sweetened', should the water quality deteriorate. This water can also be of a lower quality than that used in salmonid farming. Most surface waters are suitable although, when planning a facility, extensive tests should be undertaken with regard to quality and most importantly quantity. Water may be required all year round so summertime availability must also be determined.

If borehole or spring water is available it does have some advantages over surface supplies, especially if an intensive rearing system is planned. Ground water contains no parasites, is of consistent quality and generally has a constant temperature throughout the year. Although the temperature tends to be low (approximately 9.5–10.5°C), little water would be used during the summer, the only requirement being for topping up.

Brood fish

Good quality brood fish are one of the key requirements in successful fish rearing. These fish should be selected by body shape, size and scale pattern and be chosen in the ratio of two males to one female. Carp are generally easy to sex, especially during the spring, late summer and early autumn. The females have more rounded bellies and may have a pronounced reddened vent. The males tend to be thinner, have an inset vent and may give white milt if the flanks are gently squeezed. Brood fish should preferably be at least 4 years old, between 2 kg and 5 kg in weight and should not be 'virgin spawners', i.e. spawning for the first time. They must also be free from any visible deformities such as a shortened nose, stumpy tail or spinal flexure, as these abnormalities may be genetic in origin and could be passed onto their progeny.

Brood fish preparation is a year-round undertaking as the eggs are laid down in the summer and early autumn of the previous year. These adult spawners need to be 'pampered' in their treatment if they are to produce top quality eggs and larvae. Brood fish should therefore be kept in specially prepared ponds of at least 1.5 m depth and at low densities of between 200 and 300 kg/ha. Much effort is invested by the females in producing the egg mass which by autumn may constitute 15–20% of the total body weight of the fish. Therefore, brood fish ponds must also contain an abundance of natural food which can be encouraged by systematic manuring (see Pond Preparation later in this Section). During the summer and early autumn, the adults can also be fed a supplementary protein-rich diet (30–35% protein), at a rate of 3–5% body weight per day. Feeding can be stopped once the water temperature has dropped below 13 or 14°C. Keeping brood fish at higher densities than this can lead to egg quality problems and subsequently poor fry survival rates.

Spawning techniques

For the complete ripening of brood carp the water temperature needs to have reached and held at 18°C, preferably after a gradual warm up from winter temperatures. In the UK this is normally from the middle of May onwards although in exceptional years it may occur earlier. Optimum temperatures for the actual spawning of carp are 18–22°C.

Dubisch pond method

The simplest method for spawning carp is that developed by the Hungarian Thomas Dubisch during the second half of the last century. This system uses a small pond, usually 50–100 m^2, which has been specially constructed with drainage ditches around all four sides. The pond should be constructed in a sheltered area where it will be warmed by the sun, protected from the wind and free from disturbance. The movement of people and vehicles will disturb the carp and spawning activity can be interrupted. The pond bottom and sides are seeded with rye grass which is cut regularly. This ensures that no 'woody' vegetation or weeds become established which the carp find unsuitable as spawning substrate. In the spring the grass is cut and any dead vegetation removed. This will promote new growth which will trigger the carp spawning.

Ideally the grass should be 150 to 250 mm long and if new growth is slow it can be accelerated by covering the pond in a sheet of clear polythene. Drainage channels and catch pit can also be cleared of accumulated mud and silt, leaving the pond ready for use.

During the early part of April, the carp should be separated by sex and kept in different pools. This is to ensure that no uncontrolled breeding takes place as the survival of fry from these spawning is usually very low. When the water temperature in the holding ponds has reached and held at 18°C, the spawning ponds are filled. Dependent on the size of the pond, selected adults in the ratio of two males to one female are then introduced. The presence of suitable spawning substrate, i.e. the freshly flooded grass, stimulates the brood carp and spawning should take place within 48 hours. The carp usually spawn during the early hours of the morning attended by plenty of splashing and movement in the pond as the males chase the females. The eggs are extremely sticky and adhere to the grass on contact. Each female produces approximately 100 000 to 200 000 eggs/kg body weight and fertilization rate varies between 30% and 50%. By gently parting the grass the tiny 1.5–2.5 mm eggs should be seen. After a couple of days the clear eggs are the fertile ones and those which are milky or opaque are infertile. If spawning has not taken place within this period, the pond is emptied and the fish put back into the male and female holding tanks. The process may be repeated a few days later until spawning is achieved.

After the adults have spawned, they are removed, as their continued presence in the pond can lead to mortalities. Large fish tend to root about in the pond bottom, causing silt deposition which can smother the eggs. The adults can also eat the eggs or infect the young fry with parasites. By lowering the pond level the brood fish can be hand-netted out of the peripheral drainage ditches. This operation is best done during the early hours of the morning or on an overcast humid day, as direct sunlight on unprotected eggs can kill them. The pond can then be gently refilled with water of the same temperature and quality.

The eggs hatch after 3 or 4 days at 20°C and the larvae attach themselves to the vegetation. At this stage the newly hatched larvae are only 5 mm in length and it is important not to disturb them as they are absorbing the yolk sac and if disturbed may sink to the bottom among the roots and suffocate. After a further 2 or 3 days the larvae have absorbed almost all of the yolk. The larvae now swim up to the water surface; each fills its air bladder and begins to feed for the first time. These larvae must now be caught and transferred into the previously prepared fry ponds. This is best done by either netting them from the surface of the pond or by lowering the pond level very gently until the larvae are concentrated in the drainage ditches. They can be harvested using fine, soft mesh dip nets although great care must be taken to prevent the fry from drying out or being damaged. At this stage the larvae are very delicate and will die if handled roughly. They can then be transported in buckets and gently introduced into the fry rearing ponds.

Induced spawning

The use of pituitary preparations and various hormones for inducing carp to spawn has improved control over reproduction enormously. The subject is a complex one (see *reading list*).

Simply, the technique is to inject the brood fish with a pituitary extract or synthetic hormone when the fish is 'ripe' and ready for spawning. This mimics the natural hormone surge which takes place in the fish at this time. The hormone may be administered as a single dose or may be split up into two separate doses called the 'primer' and 'resolver', the injections usually being 12 h apart. Spawning takes place 12 to 18 h after the last resolving dose has been given. The males and female brood fish are held separately in concrete or glass fibre tanks at a temperature of 20–22°C with plenty of shade and aeration. When the fish are ready to spawn they can either be stripped by hand (Fig. 3.24), spawned in tanks using artificial spawning mats, or introduced into Dubisch ponds as already described. Hand stripping is much more labour

Fig. 3.24 Hand stripping a female chub.

intensive and expensive than tank spawning as it requires specialized equipment and the eggs have to be incubated in jars or troughs. The fertilization rate and subsequent larvae production is, however, much greater:

Hand stripping

Before the brood fish are handled they are anaesthetized using MS 222 or Benzocaine. The female is then dried carefully and gently squeezed to check whether the eggs are ready to be stripped. If only a few eggs are present at the vent the female should be returned to the tank and checked again 1 to 2 h later. If the eggs flow freely they can be stripped into a dry plastic bowl. The eggs are normally olive green in colour and approximately 1–1.5 mm in diameter. Do not squeeze the fish too much as internal damage may occur, and only continue stripping whilst the eggs are flowing freely.

The male can now be stripped and the milt poured on top of the eggs. The milt should be white, thick and creamy with no blood contaminating it. Care must be taken to prevent any urine, blood, faeces or water coming into contact with the eggs and milt as this will adversely affect the fertilization rate. The egg and milt mixture is stirred for 1 or 2 minutes after which a fertilization solution made from 40 g NaCl, 30 g carbamide or urea, and 10 litres of water is mixed in. This solution stops the eggs from clumping together and allows the eggs to swell at a slower rate. The eggs should be stirred for 40 or 50 min and the solution replaced every 10 min. After about 1 h, the eggs will have swollen to 1.5–2.5 mm and increased their volume by three to six times.

If the eggs are to undergo *jar incubation* they will need to have the stickiness washed from them. Failure to do this would result in the eggs sticking together in one mass, suffocating those eggs in the middle and making it difficult to incubate the rest. The most popular method uses a series of tannic acid washes which coagulate the remaining compounds causing the stickiness. A 0.05% solution is made by adding 5 g of tannic acid to 10 litres of water. This solution is added to the eggs and the mixture stirred for 20 s. This is then poured away and four further washes of 0.04%, 0.03%, 0.02% and 0.01% tannin solutions are given. The eggs are then washed for 5 min in fresh water, after which they are ready for incubation in the jars. At this stage the eggs are known as water hardened and are completely non-sticky.

Hatchery jars come in many shapes and sizes but are essentially inverted bottles with open tops and narrow bottoms. They are normally mounted in a row above an open trough. Water is introduced from the bottom or neck of the bottle and eggs are kept in suspension by the upwelling flow of water. The water requirement is very low, using only 0.5 to 2.5 litres/min depending on jar design. Jars normally have a 5 to 10 litre capacity although smaller jars which hold 1.5 to 3 litres are more useful for batches of eggs. In volume 1 litre represents 100 000 to 200 000 swollen or water-hardened eggs.

During the incubation period some eggs may become opaque. These infertile eggs must be removed by siphoning as they will quickly become infected with fungus which in turn causes losses among the healthy eggs. By adjusting the flow inside the jar the more buoyant dead eggs can be concentrated at the top of the egg mass and they can then be easily removed. Another method of fungal control is to turn the water supply down low and gently stir in malachite green to a concentration of 1:200 000 or 5 mg/l. This is left for 5 min, then the water supply is turned back on. This treatment is administered daily until the third day of incubation.

When the eggs begin to hatch they can be gently siphoned out of the jar and into the trough. Hatching can also take place inside the jar as the upwelling water will carry the larvae out of the bottle and into the trough below. To accelerate egg hatching the flow to the jar can be reduced to a minimum for a few minutes. This has the effect of stimulating the production of enzyme from the head gland of the larvae. This enzyme is used to dissolve the egg shell, making it easier for the larvae to hatch. Because the larvae are now in a reduced-oxygen environment their activity greatly increases, which also assists in rupturing the egg shells. As little or no flow is passing through the jar the enzyme becomes concentrated, thereby helping in the hatching of other eggs. When the water flow is turned back on, mass hatching of the eggs usually takes place in a few minutes. However, great care must be exercised using this technique as the larvae can easily be killed.

Trough incubation is an alternative to jar incubation. If there is plenty of room in the hatchery and an abundance of water, the eggs can be laid out onto 1 mm mesh trays placed inside troughs (Fig. 3.25). Water requirements for troughs vary from 2.5 to 5 1/min, depending on the trough size and number of trays. The advantage of trough incubation is that the eggs do not need the stickiness washing from them and they can be laid directly onto the trays only 30–40 min after fertilization. Also, because the stickiness has not been removed, the hatching rates tend to be higher. The compounds causing the adhesiveness of the eggs are thought to have antifungal properties and fewer eggs are lost by fungal infection. During incubation the eggs can be treated with malachite using the same method as that used for jars. When the larvae hatch they wriggle through the tray mesh and lie on the bottom of the trough. Here they stay for 2 or 3 days as they absorb the yolk sac. When the larvae have swum up, they are ready to be stocked out into the pre-prepared nursery ponds.

Tank spawning

Carp can be spawned in tanks containing artificial spawning mats known as kakabans. The males and females are kept separate until they have been injected with the final resolving dose. The two males and one female are then placed in a large, shaded, well-aerated tank containing the artificial substrate. The tank should be continuously supplied with fresh 22°C water and a strong cover must be fitted as spawning fish can jump out of the tank.

Fig. 3.25 Pouring barbel eggs onto incubation trays.

Kakabans can be constructed from 500 to 750 mm strips of greenhouse shading or soft fine mesh. These are formed into bunches, weighted with lead and distributed within the tank to form fairly dense artificial weed beds. The bottom of the tank can be covered with artificial grass which has been attached to a weighted frame. This will catch any eggs which fail to attach themselves to the kakabans.

After the adults have spawned they are removed and the eggs allowed to incubate until hatching. The tank should have a fine 1 mm mesh outlet screen to prevent the escape of swimming larvae.

First feeding

When the larvae swim up for the first time they begin to look for food. If they are not fed or stocked out within 2 or 3 days the majority of the larvae will either die or suffer long-term, indeed irreversible, damage. At this stage the gut of the larvae is being formed and it is important that food should be passing through, otherwise the gut will not develop properly. Natural food such as rotifers and protozoans or artificial diets such as egg yolk preparations or

Artemia shrimp can be used. It is important that the food size be no larger than 300 microns as first feeding larvae are unable to swallow larger items. As the fry grow the food size can be increased so that after approximately 1 month they can be fed 0.5 to 1.5 mm items such as *Daphnia*, Moina and small pellets.

Pond preparation

Nursery ponds

These are ponds in which the newly hatched carp spend the first year of their lives. Typically they range in size from 100 to 1000 m², are 0.5 to 1.0 m deep and are completely drainable. They are filled immediately eggs are seen in the Dubisch ponds. The ponds are not filled prior to this as invertebrate predators such as *Dytiscus*, *Notonecta* and *Cyclops* could become established, causing heavy losses among the fry. Cyclopoid copepods at a density of only 100/litre can kill 90% to 95% of the carp larvae within a very short time.

The ponds are then fertilized to promote the growth of phytoplankton blooms which are in turn preyed upon by the small zooplankton such as rotifers and protozoans. These small zooplankton are the ideal first food for the larvae and numbers of them quickly multiply under these conditions.

Covering the ponds with polythene clad horticultural tunnels can accelerate the response to filling and manuring. The elevated air and water temperatures inside greenhouses also increase the fry growth and survival rates. Development of the pond in terms of food size and abundance should be checked using a low-powered microscope. When plenty of rotifers and protozoans can be seen the ponds are ready to be stocked with the larvae.

Manuring rates vary, depending on soil type and whether organic or inorganic fertilizers are used. Well rotted cow manures and horse manures can be used but the best results are obtained using dried poultry waste (DPW) (see Table 3.4). This is added at a rate of 1000 to 20 000 kg/ha either by broadcasting it onto the surface of the pond, or preferably by mixing it into a slurry and pumping it in. Inorganic fertilizers can also be used and good results have been obtained using triple super phosphate at a rate of 600 to 750 kg/ha.

Table 3.4 Natural food organisms found in water and bottom soil of manured and non-manured ponds. (From Rappaport *et al.* 1977.)

Manure type	Phytoplankton (1000s/litre)	Rotifers (individuals/l)	Chironomids (individuals/0.1 m²)
Chicken manure	16.4	1000	680
Liquid cattle manure	5.6	867	163
Bedding cattle manure	3.1	247	38
Control (no manure)	2.5	170	59

Growing on ponds

The fingerling carp which were reared during the previous summer are then reared for a further 1 or 2 years in growing on ponds. These range in size from 1000 m^2 to huge ponds of 5 to 10 ha. In the UK ponds for growing on should be filled in early March. This gives 8 to 10 weeks for the mature zooplankton bloom (*Daphnia* and Copepod spp) to develop ready for the introduction of fish in early May. Filling the ponds before March may lead to problems of poor phytoplankton blooms because of low temperatures and short day lengths. As a result the production of *Daphnia* and bottom living organisms such as *Tubifex* and Chironomids would be greatly reduced.

Application rates for manuring are identical to those used in the nursery ponds although the natural bloom is allowed to commence before the fertilizer is added.

All nursery and rearing ponds would benefit from small regular introductions of manure. This is added at a rate of 20–30 kg/ha/day which will continue to promote the growth of natural food in the pond. The bacteria and protozoans which coat the manure particles are protein rich and can be directly ingested by the carp. However, if manure is to be added on a regular basis care must be taken with the introductions as too much could cause water quality problems. Conversion rates for manure introduced and fish flesh produced are approximately 4:1.

Wintering ponds

Wintering ponds are used so that nursery and rearing ponds can be drained, dried out and prepared for the next growing season. Carp can be overwintered in tanks especially if ground water is available but it is more probable that ponds will have to be used. These wintering ponds range in size from 500 to 20 000 m^2, are deeper than normal rearing ponds and must have a good supply of fresh water. They are dried out and worked upon during the summer and filled in early August. The ponds are then manured with DPW at a rate of 500–1000 kg/ha. This allows the natural food to develop ready for the introduction of fish in late October and early November.

Stocking densities and growth rates

Stocking densities and growth rates are dependent on the natural productivity of the pond, the level of supplementary feeding and the availability of oxygen. Carp can be grown under low density (extensive) conditions or high density (intensive) conditions.

Extensive rearing requires no supplementary feeding or aeration and growth is determined by the supply of natural food in the pond. In intensive rearing systems the density or biomass of fish held in the pond is too high for the available natural food. Therefore a supplementary protein or carbohydrate rich diet must be fed to give good growth rates. Oxygen will also be a limiting factor and aeration in the form of underwater air diffusers or mechanical agitation using pumps, paddle wheels, etc., must be provided.

The choice between an extensive and an intensive system will depend on the input costs and the amount of fish produced. Extensive rearing is relatively inexpensive when compared to intensive rearing as no feed and power costs are incurred. Daily labour requirements are also much lower. However, intensive systems are capable of producing 4–6 times more fish than extensive systems, which may justify the increased investment.

Stocking rates (low density)

Nursery ponds

Larvae can be stocked into nursery ponds at 20 to 30 m^2 (200 000–300 000/ha), without the need for supplementary feeding or aeration (Fig. 3.26). Prior to their first birthday, the young carp are known as C0+ and survival rates from larvae to 2.5–5 cm should be 50–60% if the ponds have been prepared properly.

Fig. 3.26 Stocking out larvae.

Rearing ponds

Fingerlings or C1+ as they are known, can be stocked into growing on ponds at a density of between 1 and 2 m^2 (10 000–20 000/ha). Survival rates for these fish up to the end of their second year of growth can be as high as 95% and in a productive pond they should reach a size of 20 cm to 25 cm and a weight of approximately 300 g to 350 g. C2+ fish can be grown at a density of between 500 and 750 fish/ha, again with very high survival rates. They should reach 35 cm to 40 cm by the end of their third summer of growth and weigh approximately 1.5 to 2 kg.

C3+ require a large amount of space and the maximum stocking density is between 50 and 75 fish/ha. At the end of their fourth summer these fish should weigh between 4 and 5 kg.

Stocking rates (high density)

Nursery ponds

If aeration and supplementary feeding are used the stocking densities can be greatly increased and larvae can be stocked at 100–120 m^2, (1 000 000–1 200 000/ha).

By the time the natural food has been exhausted in the pond the carp fry should be large enough to accept small pellets. The fish are fed *ad libitum* either by hand or preferably by automatic feeder. If uneaten food is observed feed rates should be reduced to avoid water quality problems. As the fry grow, natural food can be pumped in from other prepared ponds using slow revving pumps. This increases the growth rate and freshens the pond.

Survival and growth rates are comparable to those obtained at low density stockings.

Rearing ponds

C1+ fish grown under intensive conditions can be stocked at a rate of 10–15 m^2 (100 000– 150 000/ha). The fish are fed a supplementary diet of 30–35% protein at a rate of 4–6% body weight/day. Growth under these conditions with good aeration is extremely rapid and fish may grow to over 25 cm by the end of their second summer.

C2+ fish can be grown at 2–3 m^2 (20 000–30 000/ha) and when fed a 4–6% body ration may attain a size of over 35 cm by the end of their third summer.

It is unlikely that C3+ fish over this size would be grown on in a rearing system, although a limited number could be grown on with the C2+ fish.

Stocking rates (wintering ponds)

Stocking densities in wintering ponds should not exceed 5000 kg/ha. Water should be continuously added to the pond at a rate of 2000 to 3000/litre/h/ha and the fish should be kept as quiet as possible.

Harvesting

Fish are normally harvested from the rearing ponds during late October and early November when growth has almost ceased. The water will have dropped below 10°C, and at these temperatures the fish are able to withstand the rigours of harvesting. If possible the majority of fish should be removed by seine netting and the remainder taken out when the pond is emptied. This is done by draining the water off in a controlled manner until all the fish are concentrated in the catch pit immediately in front of the monk. If the water is allowed to drain away too quickly the fish may become stranded or damaged against the outlet screen and mortalities will occur. The fish can now be hand netted out of the pond and either sold, kept in holding tanks or transferred into the wintering ponds.

Holding facilities

After harvesting, fish may need to be maintained in tanks for a short while. These concrete or glass fibre tanks can be used to hold the fish prior to counting, grading, disease treatment and stocking out. Each should be fitted with a strong, hinged cover and have its own controllable supply of good quality oxygen-rich water. The tank must also be drainable via a central mesh screen into a common channel. Here the water can be collected, treated and re-used or it can be discharged directly to waste. Tanks with a 2–3 m^3 capacity receiving 1–2 litres/s of oxygen-saturated water can comfortably hold 100–120 kg of fish. The holding capacity can be increased by supplementary aeration using a centrally mounted blower feeding diffusers in each tank.

Pond management

Pond management and hygiene are important as they directly affect the health and growth rates of the fish. All rearing ponds should be emptied and left fallow for at least 3–4 weeks each year. Drying out the pond kills off most of the parasites and greatly improves the condition of the bottom soil. As soon as the pond is empty all the drainage ditches and monk chambers must be cleared of accumulated silt which has built up during the previous growing season. This is essential as it speeds up the drying process and allows any seepage or precipitation to quickly drain away. The drying out of the pond bottom is necessary for the following reasons:

- Exposure to the air will oxidize the bottom mud and encourage the development of bacteria which break down the organic sediment.

- Drying off prevents the build-up of dense weed growth which is a problem in ponds under cultivation all year round.
- Silt consists of between 70% and 90% water and drying off compresses the deposit, enabling it to be worked or removed.
- Some parasites and invertebrate predators are killed off.
- Muds are mineralized and nutrients that have been locked in the mud are released and can be recycled.
- Remedial work to the pond bottom can be carried out, e.g. filling in low points, repair of monks, removal of excess silt, etc.
- Exposure of the bottom to frost will break up and loosen silt by freeze thaw action.

To speed up the oxidation and improvement of the pond bottom the silt deposits are raked by hand three or four times during the period in which the pond is dry. The 'ideal' to aim for is a hard bottom with a 3–5 cm layer of well-aerated, loose, friable organic silt on top (Fig. 3.27).

After the pond has been raked and is beginning to dry out lime (CaOH) can be applied at a rate of 1500 kg/ha (Fig. 3.28). The lime disinfects the pond bottom, improves the quality of the soil and helps to eradicate pests such as

Fig. 3.27 'Ideal' pond bottom. Note outlet chambers and peripheral ditches.

Fig. 3.28 Dried and limed pond in October.

snails and beetle larvae. It is usually introduced by hand and should be spread evenly over the whole of the bottom of the pond at least 2–3 weeks before they are filled again. The catch pit immediately in front of the monk must be thoroughly covered as this area tends to be the wettest and snails may still be present. Control of this intermediate parasitic host is necessary as problems with eye fluke may occur in crowded culture conditions. Snails will repopulate the pond during the growing season; therefore annual treatments are necessary to break the infection cycle. Fig. 3.29 shows a large earth pond system.

Recirculating systems

Due to the relatively short growing period for carp in the UK the use of warm water recirculating systems has greatly increased over the last 10 years. Small systems which use very little water and energy are now capable of producing large amounts of fish. Because optimum temperature can be maintained carp grow very quickly in these systems, taking less than 6 months to reach 1 kg.

Basically a system comprises a series of rearing tanks housed in a well-insulated building. Water from the rearing tanks passes by gravity through a primary settlement filter. Here the faeces and uneaten food are allowed to settle out and are removed from the system. The partially treated water is vigorously aerated and then passed through a biological filter containing material with a large surface area. Here bacteria which coat the surface of the material convert the toxic ammonia into harmless nitrate. The treated water is then oxygenated and returned to the rearing tank. The temperature of the water is controlled directly by immersion heaters or indirectly by space heating the whole of the building. Water requirements are low and with good insulation the energy costs can be kept to a minimum.

If the system is operated at high stocking rates the ammonia levels, pH and oxygen content must be regularly checked.

3.7 Crayfish farming

The species of crayfish native to Britain is the white-clawed crayfish (*Austropotamobius pallipes*) (Fig. 3.30), but the species most commonly reared is the signal crayfish (*Pacifastacus leniusculus*) (Fig. 3.31). This is preferred as it grows faster and larger, reaching a marketable size (10 cm body length) in 3–4 years. When mature it can weigh in excess of three times as much as our native species and is reported to be resistant to (**but a carrier of**) the crayfish plague

Fig. 3.29 Calverton fish farm – a coarse fish unit in Nottinghamshire.

Fig. 3.30 Native crayfish – white clawed crayfish.

Fig. 3.31 Signal crayfish.

(*Aphanomyces astaci*) – a fungal infection which has decimated populations of native crayfish in many parts of the country.

The signal crayfish is native to western Canada and northwest USA. In Britain stock can either be obtained from established farms or through agents who import supposedly disease-free stock from Sweden. The Wildlife and

Countryside Act 1981 restricts the release of signal crayfish into the 'wild' either deliberately – in which case a licence must be obtained – or accidentally – where farmers are obliged to take all reasonable means to prevent their escape into the wild. Owing to the major decline in stocks of our native white-clawed crayfish people should be discouraged from introducing signal crayfish into or even near any catchment still containing a healthy stock of the native species. The crayfish plague is extremely virulent and is capable of spreading both downstream (very rapidly) and also upstream. The signal crayfish can supposedly breed with the white clawed crayfish, but it is reported to produce sterile offspring.

Crayfish require a high pH (over 6 and preferably above 7) to grow and reproduce. The ponds into which they are introduced should have a hard bottom with many suitable shelters and the minimum area of flat bottom. Therefore a V-shaped profile will give the greatest production potential. Water temperatures should exceed 15°C for at least 3 months of the year. Stocking can be accomplished either by using 'berried' females (carrying up to 250 fertilized eggs externally under the tail), juveniles brought in during the third stage of their lifecycle (when they are about 10 mm long and independent of their mother) or yearlings. They are omnivorous, feeding on scavenged carrion, aquatic plants and algae and even fallen leaves. To establish a good breeding population stocking should be carried out for 2 or 3 years although even a single introduction can in time give a well-structured population.

The first harvest should take place when the animals are 3 years old. Production can be as high as $1 \text{ kg/m}^2/\text{y}$, although an average pond should be expected to yield approximately 100 kg/ha. Harvesting is usually carried out with traps during the summer months between the end of the spring moult (end of May) and the start of mating (normally early October).

Crayfish have many predators, particularly when young. Prior to stocking efforts should be made to eliminate predatory fish, especially eels. However, they can be grown very successfully in angling ponds, particularly in trout lakes, although it should be remembered that they will readily take an angler's bait and can therefore be a problem in coarse fisheries where 'bottom' fishing is popular.

The Turkish crayfish (*Astacus leptodactylus*) is also common in some waters round London, where it is causing similar problems for anglers. This species is not commercially farmed (Fig. 3.32) but big populations have developed from escapees from fish markets.

3.8 Other users of the aquatic environment

Natural and artificial waters are very widely utilized for leisure activities and many fishery managers will find they have to cope with several forms of

Fig. 3.32 Turkish crayfish.

recreation in addition to fishing. A description of each will give some idea of their needs and compatibility, and how they can take place at the same time and location as fishing.

Boating

Inland pleasure cruising embraces use of the inland waterway system by a variety of craft, many of which may be provided with sleeping accommodation. A wide range of boats is used, including converted narrow boats, cabin cruisers and other craft, all of which may vary considerably in size. The boats require adequate mooring facilities and these are often provided by marinas. Bankside moorings can exclude anglers from substantial stretches of bank. Cruising can take place only on waterways such as canals or river navigations. A licence is required from the navigation authority to operate on such waters, whether the craft is powered or not. Passenger pleasure boats may operate on some rivers that are not navigations with the consent of the riparian owner.

Other boating activities include the occasional use of small boats by private individuals, or those hired by commercial undertakings. Access to the water is made via private or public slipways (Fig. 3.33) which may also provide mooring facilities.

Rowing, in an organized form, may occur on large lakes and rivers, and continues throughout the year. Most clubs organize their own regattas and head races. Rowing clubs normally own their own boats, and build or hire boathouses and club houses on the water's edge.

Fig. 3.33 A public slipway on the River Trent.

Canoeing has increased tremendously since 1950 with the advent of glass fibre construction kits, and the sport can be split into three types. Canoe touring requires unrestricted passage over long stretches of river, whereas the 'white water' canoeist needs a length of rough water, normally only to be found on fast-flowing streams and rivers or below weirs or open sluice gates. Canoe racing uses lakes or rivers for a course of 1000 m or more and is conducted rather like rowing. There are some stretches of river specifically set aside for canoeists, whereas other stretches may be used occasionally with the consent of the riparian owner.

Sailing is practised on a variety of waters, and special facilities, including club houses and slipways, are often provided on reservoirs. Sailing is normally restricted to those waters that provide sufficient space for boats to tack up-wind. On large rivers with little current, they will often be forced to approach close to the bank when sailing up-wind.

Sail-boarding is a sport that has become very popular over the last few years, and is carried out mainly on stillwaters. Because the sail-boards are usually transported to and from the water on every visit, bankside access must be good. Needs are similar to sailing.

Sub-aqua

Sub-aqua diving includes the use of underwater breathing equipment, face masks and simple swimming equipment. Diving takes place in reservoirs and lakes and on some rivers, but divers normally prefer those waters of high clarity and reasonable depth. Divers are normally content with a very limited area in these conditions.

Fig. 3.34 Swimming in a 'swimmers only' corner of a lake in Nottingham.

Swimming

Swimming is the most popular of all water-based activities (Fig. 3.34). A few rivers, gravel pits and other enclosed waters are used, but seldom provide ideal conditions. Casual swimming can take place anywhere in warm summer conditions, but it is banned under byelaw in navigations. A word of warning is appropriate: however clean a surface water looks, there is always the possibility of infection to human beings who swim there.

Water skiing

Water skiing is a recent sport which has grown to its present popularity in little more than 20 years. Certain inland waters are leased specifically for this activity, but there may also be water skiing zones on large rivers. Because of the noise over and under the water, and the wash created by the powerful boats needed to tow the skier, water skiing activities may cause conflict with other water users, especially if the skiers approach closer than 30 m from the bank.

Informal recreation

It is the quiet attractive appeal of streams, rivers and other inland waters that make them among the most valued areas for informal recreation. The popular belief in the image of cool, shady streams flowing through flowery meadows (or its upland counterpart of bubbling water dashing over rocky waterfalls) draws millions of people each year to picnic and play near water. This activity normally requires footpaths and organized or informal picnic places, which by their very nature change the environment that people come to enjoy. At worst, inconsiderate behaviour by adults and children, and lack of control of their dogs, can disrupt other activities such as angling, which require quiet.

Nature conservation

The NRA has a duty to further the conservation of rivers and to promote conservation of the aquatic environment on inland waters. Any function of the NRA has to take this duty into account. The duty also relates to the privatized water services and to internal drainage boards. The Wildlife and Countryside Act (1981) also offers protection to various species and to sites of interest for biological or geological purposes. The fishery manager should be aware of all relevant legislation relating to conservation of the natural or built environment.

There are many protected species of mammals, birds and fish (a list is given in Appendix 10) and trees may have special protection by tree preservation order (TPO).

Any work alongside a river or inland water should be checked for wildlife and historic sites. For example, a sluice may well be a listed structure and consent will be required from the local authority to undertake any work on it. A list of the various conservation designations is given in Appendix 11.

Exploitation and use

- Angling, requirements, methods, setting up an angling club
- Commercial use of coarse fish and salmonids
- Aquaculture of salmonids, cyprinids and crayfish
- Other uses

Aquatic environment

- Water quality and quantity
- Basic fish biology
- Food and food chains
- Monitoring fish stocks
- Fish mortalities

Management techniques and methods

- Management of coarse and salmonid fisheries
- Control of aquatic plants and pests and predators
- Construction and maintenance of stillwaters
- Habitat improvement in still and running waters
- Protection of stocks by regulation
- Bankside vegetation

Fig. 3.35 Summary of freshwater fisheries management.

Appendix 1
Administration of freshwater fisheries in the British Isles

All inland waters are owned by somebody, and fishing rights may be owned together with the land on which the water is situated, or separately from the ownership of the land. However, the situation varies in different parts of the British Isles, although the individual rights of an owner will be recorded in the deeds of the property. Scottish law is based on a different system from that of England and Wales, although much of it is the same and the ultimate Court of Appeal for all is the House of Lords. Northern Ireland is subject to many of the laws of England and Wales although there are local provisions. The Republic of Ireland has its own legal system.

The administration of an individual fishery will vary depending on whether it is being used for angling or for commercial fishing. This is usually left to the needs of the owner. The management operations undertaken by the fishery owner will be influenced by the common and statute law of the country and how fisheries legislation is administered at national and/or local level.

Scotland

The organization and administration of salmon and freshwater fisheries in Scotland is quite different from that in any other part of the British Isles. The right of salmon fishing (which includes sea trout fishing) derives directly from the Crown. There are no public fisheries and all salmon and sea trout fishing is privately owned, both in inland rivers and the sea out to the limit of territorial waters. Those fisheries which have been part of the subject of a grant to a member of the public in times past are heritable property and may be separated from the land in riparian ownership. (The exceptions are Shetland and Orkney.)

The main salmon and freshwater fisheries legislation in Scotland is uniquely Scottish in both origin and application. The Acts which could be cited are the Salmon Fisheries (Scotland) Acts of 1862 and 1868, the Salmon and Freshwater (Protection) (Scotland) Act 1951 and the Freshwater and Salmon Fisheries (Scotland) Act 1976 (21) and (22). The Diseases of Fish Acts 1937 and 1983, the Sea Fish (Conservation) Act 1967, the Fisheries Act 1981 and the Salmon Act 1986, which have provisions relating to salmon and freshwater fish, are Great Britain Acts and therefore apply both to Scotland and to England and Wales.

The Secretary of State for Scotland has overall responsibility for salmon and freshwater fisheries. The Freshwater Fisheries Division of the Scottish Office of Agriculture and Fisheries Department (SOFAD) is the central administrative body for salmon and freshwater fisheries in Scotland. The Department's Freshwater Fisheries Laboratory, at Pitlochry, provides information and advice on salmon and freshwater fisheries matters to the central administration.

Local administration of salmon and migratory trout fisheries together with statutory protection is carried out by Salmon District Fishery Boards. There is provision for the constitution of 108 District Fishery Boards, with a district in nearly all cases being based on the watershed of a river system and extending along the coast on either side of the river mouth. Just over half, 58, of the Districts have formed Boards.

A Fishery Board has the power to:

- Impose a fishery assessment on each salmon fishery in the district (this is their main source of income both in river and in the sea) in proportion to the value of each;
- Do such works and incur such expenses as are expedient for protecting and improving the fisheries within the district.

A District Fishery Board has no standing in respect of brown trout or other species of freshwater fish.

The main function of the seven River Purification Boards is to maintain and, where necessary, improve the quality of rivers and coastal waters. This responsibility includes investigating and monitoring all pollutants discharging into rivers or the sea. Biological monitoring is also undertaken to assess the status of plant and animal communities as an indicator of changes in water quality. The Board has the power to prosecute pollution offenders.

England and Wales

There are several organizations in England and Wales with fisheries responsibilities. The principal one is the National Rivers Authority (NRA).

Historically the fisheries function in the NRA can be traced back to the 1860s. The conservation of freshwater fisheries was made a statutory function in 1865 with responsibility vested in boards of conservators. Such boards administered fishery districts based on river systems.

The Salmon and Freshwater Fisheries Act 1923 consolidated all previous Acts, clearly defining the duties of fishery boards and their area of operations.

The River Board Act 1948 incorporated fishery boards into river boards. For the first time, fishery conservation activities were pursued alongside pollution control and land drainage activities – functions which were all essential for the proper management of river systems.

The measurement and allocation of water resources was included as an additional duty of river authorities, as directed by the Water Resources Act 1963, in 1965. The river authorities had a responsibility to manage defined functions in specified catchment areas.

Regional water authorities (RWAs) were set up in 1974. RWAs were given the clear duty to manage all aspects of the water cycle, incorporating the previous river authorities and the water supply and sewage treatment functions of local authorities. The setting up of RWAs required that the duties of 1600 authorities employing 75 000 people were incorporated.

The fisheries responsibilities of the RWAs were stated in the Salmon and Freshwater Fisheries Act 1975. RWAs were required, for the first time, to maintain, improve and develop all freshwater fisheries. Each developed fisheries policies that were designed to meet local needs.

One of the main criticisms of the RWAs was that the function responsible for protecting the aquatic environment was part of the same organization that polluted it! (The poacher and gamekeeper argument.)

The Water Act 1989 divided the functions of the ten RWAs, Water supply and sewage disposal were transferred to the water services plcs and the NRA took over the regulatory and environmental responsibilities (those of pollution control, water resources, flood defence, fisheries, recreation, conservation and navigation). The NRA is a non-departmental public body (NDPB) sponsored by the Department of the Environment. It also has links with the Welsh Office and the MAFF. These departments are responsible for government policy on flood defence and fisheries.

The NRA is partly self-financing. It recovers much of its costs from precepts and levies for flood defence works and through charges for water abstraction, land drainage consents, fishing licences and discharge consents. The remainder of its operational budget is funded through government Grant-In-Aid (GIA).

The Fisheries function (which is combined with Recreation and Conservation and Navigation where appropriate) is funded principally from the sale of fishing licences. The deficit is made up from the government GIA given to the NRA by the Treasury.

There has been a government proposal to create an environmental protection agency. Such an agency would incorporate all the NRA's functions. (A similar agency was also proposed for Scotland.)

The MAFF is a government department. It has a minister appointed by the incumbent government with responsibility for its three functions. In relation to freshwater fisheries the ministry advises the NRA. It can also influence national policy.

There are three main MAFF centres related to freshwater fisheries:

- The London office deals with proposals for variation of licence duties and objections thereto and with confirmation of varied or new fishery byelaws.
- The Lowestoft laboratory deals with any freshwater fisheries technical matters, including fish farming.
- The fish pathology laboratory, based at Weymouth, pursues a research programme related to fish diseases and provides a diagnostic service for all known types of fish disease.

The Salmon Advisory Committee was set up by fisheries ministers, during the passage of the Salmon Bill in 1986, to provide them with advice on matters relating to the management and conservation of wild salmon in Great Britain.

There are many other organizations relevant to the fisheries function in England and Wales. A few of these are listed in Appendix 4.

Northern Ireland

Fisheries in Northern Ireland, with the exception of fisheries protection and law enforcement (which are the statutory responsibility of two other organizations, see below), are the responsibility of the Department of Agriculture for Northern Ireland. Agriculture was one of the responsibilities devolved to the Northern Ireland Parliament and fisheries in the province are therefore regulated by local legislation, the Fisheries Act (Northern Ireland) 1966 which has been amended several times, most recently in 1991 by the Fisheries (Amendment) (Northern Ireland) Order 1991.

The Department of the Environment is responsible for water resources, water supply and sewage disposal, and, through its environmental protection division, it is responsible for the maintenance of satisfactory water quality in rivers and loughs, under the Water Act (Northern Ireland) 1972. It employs both the Fisheries Conservancy Board for Northern Ireland and the Foyle Fisheries Commission (see below) as agents. The water quality branch of the Department of Economic Development provides a water quality analysis service for DOENI.

The Department of Agriculture for Northern Ireland has overall responsibility for fisheries and land drainage. It has to promote actively a programme of development for public angling. It has a fisheries research laboratory carrying out general research into freshwater and marine fisheries.

Functionally the department is divided into an administrative branch (controlling policy and finance) and a professional, technical branch (covering angling development, sea fisheries and the department's fish farm).

Fish farming in the province is regulated by the Department of Agriculture. Fish culture licences are issued and grants and loans are available for the development of fish farms.

The Fisheries Conservancy Board (FCB), set up under the provisions of the

Fisheries Act (Northern Ireland) 1966, has responsibility for fisheries protection and enforcement of fisheries legislation throughout the major part of the province. The Board also has responsibilities in the field of pollution control, both under the Fisheries Acts and as agents for DOENI.

The Foyle Fisheries Commission (FFC) was set up in 1952 by the joint action of the Belfast and Dublin Governments and under the Foyle Fisheries Acts. It is responsible for fisheries and conservation and protection in the Foyle Area.

Ireland

Many bodies are engaged in administration, management, research and development in connection with inland fisheries. The functions of the various bodies are as follows:

The Department of the Marine (DOM) promotes the development of, and undertakes planning and co-ordinating activities in relation to, inland fisheries. It is responsible for the channelling of state funds to a number of other bodies concerned with such matters.

Under the 1959–64 fisheries Acts, the department exercises considerable powers over the local boards of conservators in matters of finance, employment of staff and policy. It has a responsibility for research into various aspects of inland fisheries. The Fisheries Act 1848 laid the foundation for the present system of administration of inland fisheries. All the statutes in existence in relation to fisheries up to 1959 were embodied in one Act – The Fisheries Consolidation Act 1959. The Fisheries Act 1980 established the Central Fisheries Board and the regional fisheries boards and defined their functions, which can be summarized as follows:

- To appoint staff and to generally implement the provisions of the Fisheries Acts 1959 to 1980.
- To administer the Fisheries Acts in relation to pollution and to institute proceedings against offenders of fisheries byelaws.
- To instigate public enquiries for the passing of byelaws for the improvement or control of fisheries.
- To issue fishing licences and collect licence fees and to set and collect a rate on local fisheries so that it can fund its operations for any one year.
- To ensure that any oyster and molluscan fishery in its fishery is protected.
- To encourage, promote and develop angling for salmon, trout, coarse fish and sea-fish.

The Central Fisheries Board is a semi-state body set up by the government in 1980 for the promotion and development of trout and coarse fisheries throughout the country. Its responsibilities also extend to sea angling.

The Board is empowered to acquire waters by gift, donation, fee, farm grant or other voluntary means and to hold and to develop such waters in the interest of its members and anglers generally. The Board can undertake development of any waters where the fishing is free. It can co-operate with any person or body, public or private, in the development of angling. It works in close co-operation with the regional boards, DOM, and local angling associations. It is financed mainly from grants from the DOM.

The Salmon Research Agency was set up in 1955 by Arthur Guinness, Son & Co. Ltd and the department responsible for fisheries. The Guinness sponsorship ceased in 1989 and the articles of association were amended to recognize the sole authority of the DOM. The name of the organization was changed to the Salmon Research Agency of Ireland.

The Salmon Research Agency's primary objective is to conduct and assist in conducting scientific research for the purpose of improving developing salmon and sea-trout fisheries in Ireland and about the coast of Ireland.

The agency's income is derived from the sale of smolts, consulting fees and a grant from the DOM.

The Electricity Supply Board (ESB) has developed five rivers in Ireland for hydro-electric purposes and, allied with this, it has actively engaged in the development of inland fisheries for many years. Under the Shannon Fisheries Acts of 1935 and 1938, all the freshwater fishing rights in the River Shannon, including all its tributaries, were acquired by the ESB. In addition the ESB owns the fishing rights on all its reservoirs and on certain sections the rivers (excluding the Shannon) which have been developed for hydro-electric purposes. The ESB also operates three hatcheries and rearing stations. It is also engaged in salmon farming.

Appendix 2
Careers in freshwater fisheries

The Institute of Fisheries Management provides detailed advice on careers in freshwater fisheries, and has described a simple generalized career structure, applicable to most of the major employers in this country (Fig. A2.1).

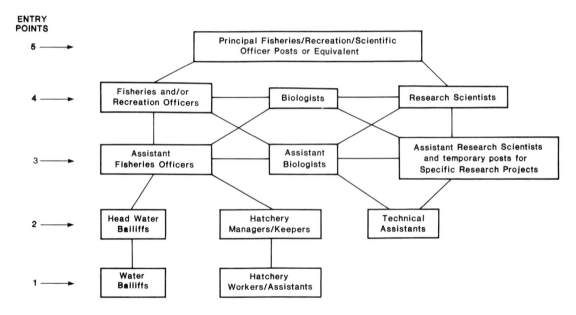

Fig. A2.1 A generalized career structure in freshwater fisheries.

Five entry points are distinguished:

Point 1: A general interest in fisheries. Previous experience of fishing methods, dealing with anglers and police work, is useful for the water bailiff positions. The Certificate of the Institute of Fisheries Management in Fisheries Management or a specialist certificate in fish farming is an advantage.

Point 2: For the head bailiff and hatchery manager/keeper positions, experience of working in a post at Point 1 is the major requirement. For the technical assistant posts, relevant GCSE certificates (or

191

equivalent) are normally required, and further qualifications are an advantage.

Point 3: Experience in a relevant post at Point 2 or a degree in a biological subject coupled with other relevant experience is expected. Post-graduate experience is particularly useful for the research posts. The Diploma in the Institute of Fisheries Management is an advantage.

Point 4: Experience in a relevant post at Point 3 would be expected, or a higher university degree in a biological subject coupled with a number of years of involvement in the fisheries field.

Point 5: A university degree coupled with several years of working in a relevant post at Point 4 is a minimum requirement.

Major employers include the NRA, MAFF, various research organizations (such as the IFE: See Appendix 3), some universities, and fish farms (although many are small).

Further training at various levels is offered by the Institute of Fisheries Management; The Hampshire College of Agriculture, Sparsholt, near Winchester; A Behrendt, Two Lakes, Romsey, Hants; Institute of Aqua-culture, University of Stirling, Stirling; Department of Applied Biology, University of Wales Institute of Science and Technology (UWIST), Cardiff; School of Maritime Studies, the University of Plymouth, Plymouth. The IFM publishes a free booklet on careers in fisheries.

Most jobs are advertized in the national press and magazines such as *Fish Farmer*.

Appendix 3
National organizations in the British Isles

The aim of this appendix is to provide an A–Z guide to the many organizations with whom a fisheries manager may come into contact, or may wish to consult.

Anglers' Co-operative Association (ACA) address: Midland Bank Chambers, Westgate, Grantham, Lincs NG21 6LE.
    The Association is concerned with handling, on behalf of its members, cases of pollution resulting in fish mortality, and with obtaining an acceptable level of compensation for members.

Angling Foundation (The) address: 23 Brighton Road, South Croydon CR2 6EA.

Atlantic Salmon Trust address: Moulin, Pitlochry, Perthshire PH16 5JQ.

British Canoe Union (BCU) address: John Dudderidge House, West Bridgford, Nottingham NG2 5AS. The BCU represents canoeists in this country.

British Waterways address: Grey Caine Rd, Watford, Herts WD2 4JR.
    The Fisheries Department of British Waterways Board is responsible for fisheries in its own canals and reservoirs. It sets the level of any rental fees, and may undertake stocking. It is very organized on an area basis.

British Water Skiing Federation (BWSF) address: 258 Main Street, East Calder, West Lothian EH5 30EE.

Central Council for Physical Recreation (CCPR) address: Francis House, Francis Street, London SW1 1DE.
    This is a national body, supported by Sports Council grant, which provides a forum for the governing bodies of sport at national level. It has six specialist divisions, one of which is water recreation.

Countryside Commission (England) address: John Dower House, Crescent Place, Cheltenham, Glos GL50 3RA.

The Commission offers expertise on all matters relating to the conservation of the landscape; provision of facilities for informal countryside recreation; the management design, and development of countryside information; and interpretation services and facilities.

The Commission was set up in 1968 under the Countryside Act. Its principal functions are aimed at the conservation and enhancement of the natural beauty of the countryside, and it encourages the provision and improvement of facilities for public enjoyment of the countryside and of open air recreation. In Wales and Scotland it is combined with the previous Nature Conservancy Council under the names **Countryside Council for Wales** and **Scottish Natural Heritage**.

Country Landowners' Association (CLA) address: 16 Belgrave Square, London SW1X 8PO.

This is the central association of country landowners, many of whom are also riparian owners of fisheries and have a keen interest in the maintenance and improvement of fisheries generally. The Association is concerned to protect lands and fisheries against harmful developments and to achieve the best deal possible for members when such projects as new trunk roads, *etc.*, are being considered at national or local level.

Department of Agriculture, Northern Ireland address: Hut 5, Castle Ground, Stormont, Belfast.

Department of the Marine address: Leeson Lane, Dublin 2.

Development Board for Rural Wales address: Ladywell House, Newtown, Powys.

English Nature address: Northminster House, Peterborough PE1 1UA.

This council is financed through the Department of the Environment and its principal functions are the establishment and management of nature reserves in England; the provision of advice to Government ministers and others; the dissemination of information about nature conservation; and the commissioning of relevant research. It is the statutory body responsible for nature conservation in Britain.

Farming and Wildlife Advisory Group (FWAG) address: Stoneleigh, Kenilworth, Warks.

Fisheries Conservancy Board address: 1 Mahire Road, Portadown, Co. Armagh.

Forestry Authority address: 21 Corstorphine Road, Edinburgh EH12 7AT.

This is the organization responsible for the promotion of forestry resources generally, including afforestation development, the establishment and maintenance of adequate reserves of growing trees, and thus the production and supply of timber.

Foyle Fisheries Commission address: 8 Victoria Road, Londonderry FWAG, NI.

Highlands and Islands Development Board address: Bridge House, Bank Street, Inverness, Scotland.

Inland Waterways Amenity Advisory Council address: 122 Cleveland Street, London W1P 5DN.

The Council is a statutory body set up in 1968. It represents the interests of all types of users on the canal and river network of the BWB.

Institute of Arable Crops Research address: Aquatic Weeds Research Unit, Sonning Aquatic Research Centre, Broadmoor Lane, Sonning-on-Thames, Reading. This has replaced the Weed Research Organization.

Institute of Biology address: 20 Queensberry Place, London SW7 2DZ.

This is a professional institute for all types of biologists. It has comprehensive training courses leading up to a degree and other levels of qualification.

Institute of Fisheries Management (IFM) address: 22 Rushworth Avenue, West Bridgford, Nottingham NG2 7LF.

This is the professional body of fisheries management. Its aims are to promote the profession and to improve the status and efficiency of its members. To this end it operates a training course leading to professional qualifications at intermediate and higher level. The Institute is organized into local branches which roughly correspond with NRA regions.

Institute of Freshwater Ecology (IFE) address: Ferry House, Far Sawrey, Ambleside, Cumbria.

The Institute conducts research into all aspects of freshwater biology, and has performed many short- and long-term studies on various species of freshwater fish and their food. The main laboratory is on Lake Windermere, with a river laboratory at East Stoke, Dorset.

Institute of Water and Environmental Management address: 15 John Street, London WC1N 2EB.

This is the professional body of water pollution control and environmental managers.

Ministry of Agriculture, Fisheries and Food (MAFF) address: Nobel House, 17 Smith Square, London SW1P 3JR.

Various branches of the Ministry are involved in different aspects of fisheries. At Great Westminster House, Horseferry Road, London SW1, are located the sections that deal with proposals for variation of licence duties and objections thereto, and with confirmation of varied, or new fishery byelaws.

At Lookout House, The Nothe, Weymouth, Dorset, is the Ministry's fish diseases laboratory. The laboratory is equipped to identify all known types of fish disease of a transmissible nature, whether bacterial or viral, and carries out research in addition to providing a pathological examination service to authorities. The MAFF laboratory at Lowestoft has overall responsibility for all freshwater and marine operations. The Director of Fisheries Research is based there.

National Association of Specialist Anglers (NASA) address: 5 Silvercourt, High Street, Brownhills, Walsall, West Midlands.

The Association exists to protect and promote the interests of those anglers seeking larger than average fish. Membership includes individuals, specimen groups and the 'single species' groups (*e.g.* Pike Anglers Club). It organizes the Annual British Angling Conference.

National Commercial Salmon Netsmen's Circle of the **Fisheries Organization Society Ltd** address: New Fish Quay, Brixham, Devon.

This is the central association of salmon netsmen in Britain, and it is concerned with all matters which may affect the interests of netting on rivers and estuaries.

National Farmers' Union (NFU) address: Agriculture House, Knightsbridge, London SW1X 7NJ.

This Union was set up in 1908. Its basic role is described as to provide the circumstances in which British farmers and growers are able to pursue their business effectively, on fair terms, and without undue pressure or interference. It represents 85% of full-time farmers in England and Wales, and has an active fish farming section.

National Federation of Anglers (NFA) address: Halliday House, 2 Wilson Street, Derby DE1 1PG.

This is the governing body of coarse angling in Britain, and deals with all aspects of the sport. It also organizes the running, annually, of the National

Angling Championship in the several divisions into which affiliated clubs are divided. It maintains contact with Central Government.

Natural Environment Research Council (NERC) address: Polaris House, North Star Avenue, Swindon, Wilts 5NZ 1EV.

NERC is responsible for encouraging, planning, funding and executing research in those physical and biological sciences that relate to the natural environment and its resources.

National Rivers Authority (NRA) see Figure 1.8 and below for regional addresses.

Office Addresses & Telephone Numbers

Head Office – Bristol
Rivers House
Waterside Drive
Aztec West
Almondsbury
Bristol BS12 4UD

Tel: 0454 624400
Fax: 0454 624409

Anglian Region
Kingfisher House
Goldhay Way
Orton Goldhay
Peterborough PE2 0ZR

Tel: 0733 371811
Fax: 0733 231840

North West Region
Richard Fairclough House
Knutsford Road
Warrington

Tel: 0925 53999
Fax: 0925 415961

Southern Region
Guildbourne House
Chatsworth Road
Worthing
West Sussex BN11 1LD

Tel: 0903 820692
Fax: 0903 821832

Thames Region
Kings Meadow House
Kings Meadow Road
Reading RG1 8DQ

Tel: 0734 535000
Fax: 0734 500388

Head Office – London
Eastbury House
30–34 Albert Embankment
London SE1 7TL

Tel: 071 820 0101
Fax: 071 820 1603

Severn Trent Region
Sapphire East
550 Streetsbrook Road
Solihull
West Midlands B91 1QT

Tel: 021 711 2324
Fax: 021 711 5824

South Western Region
Manley House
Kestrel Way
Exeter
Devon EX2 7LQ

Tel: 0392 444000
Fax: 0392 444238

Welsh/Cymru Region
Rivers House/Plas yr Afon
St Mellons Business Park
St Mellons
Cardiff CF3 0EG

Tel: 0222 770088
Fax: 0222 798555

Northumbria and Yorkshire Region
Rivers House
21 Park Square South
Leeds LS1 2GQ

Tel: 0532 440191
Fax: 0532 461889

Royal Society for Nature Conservation (RSNC) address: The Green, Witham Park, Waterside South, Lincoln LN5 7JR.

Royal Society for the Protection of Birds (RSPB) address: The Lodge, Sandy, Bedfordshire SG19 2DL.

Royal Yachting Association (RYA) address: RYA House, Romsey Road, Eastleigh, Hampshire SO5 4YA.

Salmon and Trout Association (STA) address: Fishmongers' Hall, London Bridge, London EC4R 9EL.
  The Association is concerned with the well-being of game fishing throughout Britain, and maintains contact with central government on matters affecting the sport directly or through proposed legislation.

Scottish Office Agriculture and Fisheries Department address: Pentland House, 47 Robbs Loan, Edinburgh.

Sports Council address: 16 Upper Woburn Place, London WC1 0QP
  This is an independent body which was established in 1972 by Royal Charter. It has wide executive powers and overall responsibility for matters concerning British sport. Besides dispensing grant aid, it has extensive promotional and advisory functions.

Welsh Office Agriculture Department address: Cathays Park, Cardiff CF1 3NJ.

Appendix 4
Fisheries grants

Various organizations provide grants for fishery development operations. The following is a list of organizations which angling club secretaries might wish to approach.

Sports Council

Details of Sports Council grants can be obtained from the local divisional offices. Grants or interest-free loans may be offered towards the reasonable capital cost of providing essential facilities for those taking an active part in sport: purchase of fishing rights is one example. A grant may be given of up to 50% of the approved cost of the project as valued by the District Valuer.

Countryside Commission

Details of grants for private individuals and bodies are set out in a leaflet by the Commission (CCP 79). Grants can be made in respect of any expenditure which the Commission considers conducive to the conservation and enhancement of the natural beauty of the countryside, and the provision and improvement of facilities for the enjoyment of the countryside and for open-air recreation.

Grants for fish farming

Grants are available for the commercial production of freshwater fish for food under the Farm and Horticultural Development Scheme (FHDS), but only those who are already in business in farming or horticulture may apply. Further details can be obtained from the advisers at the MAFF divisional offices.

For those ineligible for FHDS, grant aid may be available from the EEC Agricultural Fund (FEOGA). This grant is mainly for co-operative proposals between several fish farms on projects exceeding £75 000 in total (1981 baseline). Applications should again be made via MAFF.

Finally, a selective assistance scheme is available from the Department of Industry only for farms in development areas ineligible for the FHDS grant.

The Development Board for Rural Wales and the Highlands and Islands Development Board also provide grants for fish farming.

Tree planting grants

Whether or not grants are available depends very much on the scale of the planting scheme. Grants are not likely to be made for a handful of trees, but if the area involved is large enough, or if the planting can be incorporated into a larger scheme, then grants may well be payable. The Forestry Authority or your local county council should be contacted for further information. Before planting trees around your pool or alongside the river, make sure that the riparian owner is agreeable and – if it is a watercourse – check that the Flood Defence Section of the NRA has no objection. The department may require one tree-free bank from which to carry out maintenance work, and their bye-laws commonly regulate tree planting near watercourses.

Appendix 5
Checklist for the construction, restoration and management of pond fisheries

This appendix brings together information from various sections of this book to enable a fisheries manager to design, construct and manage a pond fishery. It is not unusual for fisheries managers to plan and design ponds and then undertake the construction work themselves. Others may employ contractors to undertake some or all of the work. These notes and comments should be of help to both types of manager.

Notes – facts	Comment – action
Site requirements	
Don't choose a site with high wildlife or archaeological value.	Contact county archaeologist, local FWAG officer, NRA, County Wildlife Trust.
Need water supply all year.	Contact NRA fisheries officer.
Need good water quality all year round.	Contact NRA fisheries officer.
Need soil suitable for water retention.	Either use soil auger over all site or employ contractor who will do this as part of the construction work. It is normal practice to ask three contractors to give a quotation for such work. One of these is then chosen to undertake the contract.
Due care should be paid to landscape.	Ensure that the pond is blended in with the local landscape.
The local authority requires planning permission for ponds in the following situations:	Non-agricultural use, e.g. angling, ponds within 25 m of a classified road, exporting material off site, e.g. soil and gravel.

Continued

Notes – facts	Comment – action

Type of fishery

Trout and cyprinid (coarse fish) are the two basic pond fisheries in this country.	If the ponds are to be managed as commercial fisheries then various financial constraints will have to be considered, e.g. initial and follow-up stocking costs, space required for each angler, possible cost of day or season ticket. Overall economics usually favour coarse fish, depending on local supply and demand.
The type of fishery required will affect the pond design.	Generally trout fisheries need cleaner, deeper water than cyprinid fisheries.
Large trout and carp are more likely to be stolen.	Can the site be made secure?

Water supply

The site for the new pond must receive sufficient water throughout the year, if it is to be a successful fishery. In the ideal world input must at least balance output, both in the outflow, by evaporation and for seepage.	Water supply may be from surface water, ground water or spring water. These should be assessed in the field by looking at the water table and vegetation. The NRA could have office records of the area.

Engineering aspects

There are three basic types of lakes: water table; impounding lake (dam across a stream); offstream lake.	Contact a consultant or contractor. Ask NRA for list. Generally an on-line lake is the least suitable.
There are two basic methods of construction: cut and fill (dig a hole and fill with water); total excavation, fill with water. The embankments and slopes must be properly designed and constructed.	Contact a consultant or contractor. Ask NRA for list.

Water legislation and licensing for pond fisheries

A licence is required to both abstract and impede the flow of any watercourse.	Contact the NRA to find out how to apply for a licence and details of costs.

Continued

Notes – facts	Comment – action
Abstraction licences are normally required to abstract water for most purposes.	The NRA may normally stipulate that the initial filling of a pond is restricted to the winter months (Oct to March inclusive), depending on the catchment.
Impounding licences. (Water Resources Act 1991) Might also need a consent under the Salmon and Freshwater Fisheries Act 1975 if a fish pass is needed in the weir or dam.	Offstream ponds which include works for diverting the flow of water by means of a weir or dam will require an impounding licence.
The Reservoir Safety Act 1975 applies if a pond contains more than 25 000 m^3 of water above the natural level of part of the land.	Contact the local authority in England and Wales. It is normally better to create two or more smaller pools than one large one.
The Water Resources Act 1991 Section 109, the Land Drainage Act 1991 Section 23 and Land Drainage Byelaws define the NRA's powers in relation to flood defence and land drainage.	Consent is required from the NRA for the erection, construction or alteration of any of the following works: • A mill dam or other like obstruction to the flow in a watercourse; • A culvert; • Any other structure including angler platforms over or under a main river* • Any structure or building on or near the bank of a main river. • A small charge is made for each consent.

Water quality

Water quality is determined by a number of factors (See page 5); a few may be adjusted to suit the fishery required.	For most it is easier to adjust the fishery to suit the quality of water available. It is important to check the range of conditions suitable for fish and to check the quality of the intended water supply to a pond.

* A 'main river map' means a map relating to the area of a regional flood defence committee which shows by a distinctive colour the extent to which any watercourse in that area is treated as a main river and indicates which water courses are designated in a special drainage scheme. Main river maps are to be conclusive evidence, for the purposes of the Land Drainage Act 1991, and all other purposes, of what is a main river. *(see reading list,* Howarth (1992)).

Continued

Notes – facts	Comment – action
	Contact the NRA who can arrange to take a water quality sample at an appropriate time. A small charge may be made. A fisheries consultant may also offer the same service. Books in the reading list should also be referred to. Remember: water quality varies both with time of day and with season.

Excavation methods

The stage will come, sooner or later, when construction plans have been found to be satisfactory and all legal requirements have been satisfied.	It is important to be aware of the safety aspects of excavating ponds and lakes. There are basically seven stages to follow when undertaking one of the three types of construction: • Make sure there are no public utilities in the area; • Strip and heap the soil for future use; • Excavate to the required dimensions, check for any drain and remove if necessary; • Excavate and trim the banks to a minimum of 50° from the vertical axis, with differing profiles to create different habitats; • Spread and contour the excavated material, known as spoil – you'll have consulted the NRA on how this should be done; • In the case of a dry pond construct an outlet, then an inlet system; • Finally spread the original top soil over the disturbed area.

Stocking and management

The stocking rates for coarse or salmonid fisheries vary depending on the type of fishery.	See page 153 and 171. Newly dug ponds take some time to mature, usually a spring and a summer.

Continued

Notes – facts	Comment – action
Most still water coarse fisheries will support between 300 and 400 kg/ha of fish. Fisheries range from general mixed fisheries to specialist carp or pike fisheries. General mixed fisheries – a balance should be achieved between bottom feeding species and more general feeders. In such a fishery it is advisable to stock no more than 12 carp/ha. Carp are a dominant bottom-feeding species. A good mix of other species would be roach (or rudd) crucian carp and perhaps perch. Bream should only be stocked in waters greater than 1 ha.	Contact the NRA, Fisheries Office for more advice on stocking rates. The NRA, FO will also explain the legal requirement for any person to obtain stocking consent before introducing any fish.
Crucian carp may not perform well when stocked with other species. Though they can be stocked in carp fisheries, they rarely do well as they appear to suffer from competition with the non-crucian carp.	
Overstocked carp fisheries are popular as day ticket fisheries, stocked at the rate of 1000/kg/ha. Such waters are dependent on supplementary feeding, and often experience problems. Specialist carp (ranging in size from 4 kg to 10 kg in a minimum size water of 1.5 ha) and pike (a mixture of 8:1 of prey fish such as perch, roach, bream and a few gudgeon and pike at a stocking level of 300 kg/ha) fisheries.	
The stocking rate in a put-and-take trout fishery is 75–100 kg/ha, with fish stocked in the spring and summer. Rainbow trout give a higher catch rate than brown trout, but are more likely to die over winter. Catch and release trout fisheries can give a return on stock fish of 250%	

Continued

Notes – facts	Comment – action
Old established pools decline in productivity over the years. Productivity can be restored by using hydrated lime, crushed limestone, basic slag, triple superphosphate or well rotted farmyard manure.	Contact the NRA, FO. Also English Nature to check whether the site is an SSSI.
The presence of dead fish in any water can be indicative of many causes.	Contact immediately the NRA emergency telephone number on your national fishing licence. The NRA will investigate straight away.

Aquatic plants

A freshly dug pond will need to be planted with appropriate aquatic plants (see page 89). The fisheries manager will need to choose correct plants and perhaps sooner or later be aware of the various techniques and legal requirements to control any excess growth (see page 78 and Appendix 6) Beware of introducing the wrong plants which may get out of hand and deoxygenate the water at night. This applies to all plants including the so called 'oxygenators'.	Contact the NRA, FO and the NRA Conservation Officer, the County Naturalist and the local FWAG Officer. (Note that the Wildlife and Countryside Act 1981 prohibits the taking of any wild plant without the owner's permission. Certain plants are protected totally.) The use of aquatic herbicide to control plant growth legally comes under the Pesticides Regulations 1986. The Regional Pollution Control Dept of the NRA must be notified of the proposal.

Monitoring and control of fish stocks

It is sometimes necessary to examine the fish stocks in a pond or remove the excess stocks or unwanted species from one water to another. (see Section 1.6 and the *reading list*).	See the NRA, FO.

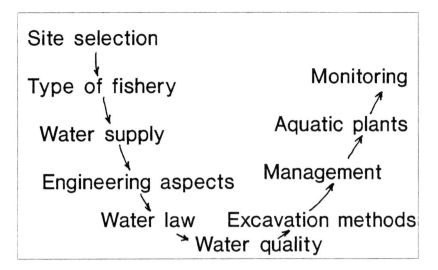

Fig. A5.1 Checklist for construction or creation of pond fisheries.

Appendix 6
Aquatic plants

Ponds and watercourses require plants both on the banks and in the water; they provide cover for animals and birds and provide cover and spawning areas for fish. Aquatic plants can be divided into four categories:

- *Marginal:* growing on the banks or waters edge but requiring a damp environment;
- *Emergent:* plants rooted in water but leaves emerge;
- *Floating:* free floating or rooted plants with floating leaves;
- *Submerged:* plant totally submerged in water.

The following list of plants is divided into the above categories and incorporates a tick list of various features.

Key
St = stillwater   Fl = flowing   Inv = invasive   H = herbicide   F = flowering attractively   P = poisonous   Ha = specific habitat for breeding birds

y = yes   n = no   s = sometimes

Herbicides

The Control of Pesticides Regulations 1986 must be followed when using a herbicide. The following list comprises the only herbicides that can be used on or near water. Fossamine ammonium and Maleic hydrazide may also be used for control of some plants on river banks. The NRA must be consulted before use of any herbicide. A COSHH assessment on the use of a herbicide must be carried out before use. If in doubt use a certificated contractor to carry out the work.

Listed by active ingredient

a   Glyphosate
b   Dalapon
c   2 Diquat
d   4–D amine
e   Terbutryne
f   Dichlobenil
g   Asulam

Plant name	Common name	St	Fl	Inv	H	F	P	Ha
Marginal								
Acorus calamus	Sweet flag	y	y	n	a	n		
Carex spp.	Sedge	y	y	n	a,b	n		y
Eleocharis spp.	Spike rush	y	y	n	a	n		
Epilobium hirsutum	Great hairy willowherb	y	y	s	d	y		
Filipendula ulmaria	Meadow sweet	y	y	n	a	y		
Glyceria maxima	Reed sweet grass	y	y	y	a	n		
Iris pseudacorus	Yellow iris	y	y	n	(a)	y	s	
Juncus effusus	Soft rush	y	y	n	a,d	n		
Lycopus europaeus	Gypsywort	y	y	n	a	y		
Lythrum salicaria	Purple loosestrife	y	y	n	d	y		
Mentha aquatica	Mint	y	y	n	d	y		
Myosotis spp.	Water forget me not	y	y	n	a,f	y		
Phalaris arundinacea	Reed canary grass	y	y	n	a,b	n		
Phragmites australis	Common reed	y	y	n	a,b	n		y
Pteridium aquilinum	Bracken	y	s	s	a,g	n	y	
Rhizoclonium hieroglyphicum	Water dock	y	y	n	e,(c)	n		
Rumex spp.	Dock	y	y	n	e,g	n		
Scirpus lacustris	Club rush	y	y	s	a	n		
Sparganium erectum	Bur reed	y	y	s	a,c,(b)	n		
Typha latifolia	Reed mace	y	y	y	a,b	y		
Typha angustifolia	Narrow leaved	y	s	n	a,b	y		
Veronica spp.	Brooklime	y	y	n	a	y		
Emergent								
Alisma plantago- aquatica	Water plantain	y	n	n	a,c,d	y		
Apium nodiflorum	Fools water cress	y	s	s	(a),(d)	n		
Berula erecta	Lesser water parsnip	y	y	n	(f)	y		
Butomus umbellatus	Flowering rush	y	y	n	a	y		
Caltha palustris	Marsh marigold	y	y	n	(a)	y		
Equisetum spp.	Horsetail	y	s	s	a,f,(d)	n		
Hippurus vulgaris	Mares tail	y	n	s	f	n		
Menyanthes trifoliata	Bogbean	y	n	s	f	n		

Continued

Plant name	Common name	St	Fl	Inv	H	F	P	Ha
Oenanthe spp.	Water dropwort	y	s	n	f	y	y	
Rorippa nasturtium-aqua.	Water cress	y	y	s	d,(f)	n		
Rorippa amphibia	Great yellow cress	y	s	n	d	y		
Sagittaria sagittifolia	Arrowhead	y	y	n	a,f	y		
Scrophularia spp.	Water figwort	y	y	n	a,f	y		
Floating								
Azolla filiculoides	Water fern	y	y	y	c	n		
Glyceria fluitans	Floating sweet grass	y	y	n	c,f	n		
Hydrocharis morsus	Frogbit	y	s	s	c,f	y		
Lemna spp.	Duckweed	y	y	y	c,e	n		
Nuphar lutea	Yellow water lily	y	y	s	a,(d)	y		
Nymphaea alba	White water lily	y	n	n	a,(d)	y		
Nyphoides peltata	Fringed water lily	y	n	s	(d)	y		
Polygonum amphibium	Amphibious bistort	y	s	s	e	y		
Potamogeton spp.	Pondweed	y	s	s	c,e,f	n		
Ranunculus spp.	Water crowfoot	y	y	n	c,e,f	y		
Stratiotes aloides	Water soldier	y	n	s	f	y		
Submerged								
Callitriche spp.	Water starwort	y	y	n	c,e,f	n		
Ceratophyllum demersum	Hornwort	y	s	s	c,e,f	n		
Cladophera glomerata	Blanket weed (algae)	y	y	y	e,(c)	n		
Chara spp	Stonewort	y	y	n	e,f	n		
Eleogiton fluitans	Floating club rush	y	s	n	c,f	n		
Elodea canadensis	Canadian pondweed	y	s	y	c,e,f	n		
Enteromorpha intestinalis	Blanket weed (algae)	y	s	y	e,(c)	n		
Fontinalis spp.	Moss	y	y	n	f	n		
Hottonia palustris	Water violet	y	n	n	e,f,(c)	y		
Juncus bulbosis	Rush	y	n	n	f,(c)	n		
Myriophyllum spp.	Water milfoil	y	y	n	c,e,f	n		

Continued

Plant name	Common name	St	Fl	Inv	H	F	P	Ha
Spirogyra spp.	Algae	y	s	y	e,(c)	n		
Utricularia spp.	Bladderwort	y	n	n	e	n		
Zannichellia	Horned							
palustris	Pondweed	y	y	n	c,f	n		

Appendix 7
Legislation for angling club administrators

It is outside the scope of this appendix to attempt to describe the role and function of the many government departments and authorities who control or operate the following Acts of Parliament. It is therefore useful to include on the club committee, a person or persons with some expertise in financial and legal matters. The following notes give just the briefest outline of the essential elements of some of the relevant statutes. There are several good books on the law in the Reading List.

Salmon and Freshwater Fisheries Act 1975

This deals with the fishery functions of the NRA. It also consolidated into one statute the previous six Salmon and Freshwater Fisheries Acts.

The Act, in addition to those duties described in Appendix 1, also controls the methods of fishing, the close seasons during which fishing may not take place, and in certain circumstances the size of fish which may be lawfully caught. It covers the licensing of users, the administration and reinforcement of the Act, and movement and introduction of fish into inland waters. It also gives the NRA powers to make byelaws.

Diseases of Fish Acts 1937 and 1983

These Acts enable MAFF to take measures to control the spread of disease by making certain diseases 'notifiable'. The Acts also control the importation of live fish and the eggs of fish.

Where the NRA suspects that any waters are infected with any disease of fish to which the Acts apply, it must report it to the Minister. In an 'infected area', the Minister may authorize the NRA to remove dying or dead fish from that 'infected area', but this does not apply to fish farms. The Acts do not enable the Minister to pay compensation as would be the case in some diseases of cattle requiring animals to be slaughtered.

Theft Act 1968

Schedule 1, paragraph 2 of the Theft Act 1968, deals specifically with taking or destroying fish from private property.

(1) Subject to paragraph (2) below, a person who unlawfully takes or destroys, or attempts to take or destroy, any fish in water which is private property or in which there is any private right of fishery is liable on summary conviction to a fine not exceeding £50 or, for an offence committed after a previous conviction for an offence under this paragraph, to imprisonment for a term not exceeding three months or to a fine not exceeding £400 or to both.

(2) Paragraph (1) above does not apply to taking or destroying fish by angling in the daytime (*i.e.* in the period beginning one hour before sunrise and ending one hour after sunset), but a person who by angling in the daytime unlawfully takes or destroys, or attempts to take or destroy, any fish in water which is private property or in which there is any private right of fishery is liable on summary conviction to a fine not exceeding £50.

(3) The court by which a person is convicted of an offence under this provision may order the forfeiture of anything which, at the time of the offence, he had with him for use in taking or destroying fish.

(4) Any person may arrest without warrant anyone who is, or whom he, with reasonable cause, suspects to be, committing an offence under paragraph (1) above, and may seize from any person who is, or whom he, with reasonable cause, suspects to be, committing any offence under this provision anything which on that person's conviction of the offence would be liable to be forfeited under paragraph (3) above.

Under paragraph (2) the maximum penalty is £200. Angling at night, and taking or attempting to take and destroy fish by any method (other than angling during the day) during the night or during the day, the maximum penalty is £1000 and/or imprisonment for a period not exceeding 3 months. The power of arrest without warrant does not apply to angling during the day time. The Theft Act 1968 stopped the 'apparent right' of owners, club bailiffs and gamekeepers from confiscating fishing tackle instead of going to court.

Reservoir Safety Act 1975

The Reservoir Safety Act 1975 has powers to make regulations and orders.

A large reservoir, any part of which is above the level of the adjacent ground and which has a capacity for holding in excess of $25\,000\text{m}^3$, will require periodic inspection by a qualified (Panel 1) civil engineer. A qualified civil engineer must also design and supervise the construction or enlargement of a reservoir, which can only then be filled in accordance with his certificate. The Act is supervised by the local authority.

National Parks and Access to the Countryside Act 1949 and Countryside Act 1968

This gives power to the local Planning Authority to enforce access to a public path and to stop notices deterring the public from using footpaths. It also gives the Authority powers of entry upon land for the purpose of surveying the land.

Section 119 of the **Highways Act 1959** refers to ploughing of footpaths or bridgeways, and gives the duty of enforcing the provisions under the Act to the Highway Authority.

Control of Pollution Act 1974

The control and prevention of pollution of rivers is the responsibility of the NRA. Most of the provisions of this Act have been re-enacted in the Environmental Protection Act of 1990 and in the Water Resources Act 1991, although some provisions still exist.

Protection of Birds Act 1954 and 1976 (Incorporated into Wildlife and Countryside Act 1981)

The 1954 Act sets out in four schedules: (1) birds which are protected at all times and during the close season; (2) wild birds which may be killed or taken at any time by authorized persons; (3) wild birds which may be killed or taken outside the close season; and (4) wild birds which may not be sold alive unless close-ringed and bred in captivity.

Section 8 of the 1976 Act would appear to allow a licence to kill birds to prevent serious damage to property or to fisheries.

Water Resources Act 1963

The Act is to promote measures for the conservation, redistributing, augmenting and securing the proper use of water resources or of transferring those resources to another area. A person who abstracts water from a watercourse or underground strata requires a licence to abstract. The construction or alteration of impounding works in a watercourse also requires a licence. The owner of a fishery can, under certain circumstances, apply for an existing licence to be revoked where he can prove damage or loss due to an existing abstraction. The erection of a culvert in a watercourse, or the alteration or removal of any mill dam, weir or other like obstruction requires the consent of the NRA. (This Act was repealed in parts by the Water Act 1989 and consolidated by the Water Resources Act (1991).)

Health and Safety at Work Act 1974

It is the duty of every employer to ensure, so far as is reasonably practicable, the health, safety and welfare at work of all his employees. He must provide and maintain a working environment for his employees that is, so far as is reasonably practicable, safe, without risks to health, and adequate as regards facilities and arrangements for their welfare at work.

The Health and Safety Executive issue many guidance leaflets, and enquiries should be made direct to Baynards House, 1 Chepstow Place, Westbourne Grove, London W2.

Sex Discrimination Act 1975

This Act is designed to promote equality of opportunity between men and women, and to render certain kinds of sex discrimination unlawful. It is largely enforceable by taking civil proceedings. The fisheries administration must be aware of the above statute prior to advertising staff.

Wildlife and Countryside Act 1981

In England and Wales the release into the wild, whether deliberately or unintentionally, of any fish or shellfish not ordinarily resident in Great Britain, or which is listed in Schedule 9 to the WLCA 1981, or the eggs of such fish or shellfish, is an offence unless the release has been authorized by an individual licence issued by the MAFF or the Welsh Office Agriculture Department. An individual licence, for which a fee will be payable, will cover the release of a single consignment into one location and will be granted if, after consultation with English Nature and the NRA or the Countryside Council for Wales, the Minister or Secretary of State for Wales is satisfied that the introduction is justified. Releases of fish into totally enclosed ornamental lakes and ponds and fish farms are outside the scope of the Act, although the permission of the NRA is still required.

The aim of this licensing scheme is to control the release of new or non-native species of fish (including shellfish) into the wild because of their effect on the indigenous flora and fauna. The fish that are established in the wild in some parts of the country, and listed in Schedule 9 to the Act (see also Appendix 10), are large-mouthed black bass, rock bass, bitterling, pumpkin-seed (otherwise known as sun-fish or pond-perch), wels (otherwise known as European catfish) and zander.

Details and application forms are available from the MAFF, Nobel House, 17 Smith Square, London SW1P 3JR or, for the NRA Welsh Region, the Welsh Office Agriculture Department, Cathays Park, Cardiff CF1 3NQ.

Section 14 of the Act prohibits the introduction into the wild of animals, including fish and shellfish, which are not ordinarily resident in and are not regular visitors to Great Britain, while section 16 provides an exemption to section 14 if a licence is granted by the appropriate authority.

Under the Act the appropriate authority is:

(1) in relation to fish and shellfish in England, the Minister of Agriculture, Fisheries and Food;

(2) in the case of salmon and freshwater fish in the area of the Welsh NRA region, and that part of the area of the Severn–Trent Region of the NRA in Wales, and sea fish and shellfish off the coast of Wales, the Secretary of State for Wales;

(3) in the case of fish and shellfish in Scottish waters, the Secretary of State for Scotland.

These provisions of the Wildlife and Countryside Act 1981 do not extend to Northern Ireland.

Police and Criminal Evidence Act 1984 (PACE)

This sets out the procedure and practices to be used in the investigation of offences and the arrest, interrogation, etc. of suspects. It also strengthens the hand of the NRA bailiffs in dealing with offenders whose identity is in doubt.

Occupiers Liability Act 1957

The occupier of premises owes to all visitors the duty to take such care as is necessary to see that the visitor (invitees and licensees) will be reasonably safe in using the premises for the purpose for which he is invited or permitted to be there. In some cases this may even apply to trespassers (*see below*).

The occupier has a duty to ensure that the condition of the premises is reasonably safe and a duty to ensure no dangerous activity is carried on in his premises.

Trespass

A person who enters upon another's land without his consent or acquiescence, or without lawful authority, is trespassing. Trespass to land is interference with the possession of land. Where land and water is leased to an angling organization, it is the leasing anglers that have possession, not the owner of the land. It is not necessary for the plaintiff to show actual damage in order to commence proceedings in a civil court. It is possible to obtain an injunction without proof of damage.

It is suggested that if a trespasser is found, he should first be made aware of his trespass; he must then be asked to depart peacefully, and be given time in which to quit the land.

There are many other Acts of Parliament and Common Law rights, and the fisheries manager is advised *always* to seek legal clarification.

Water Resources Act 1991

This consolidates provisions contained *inter alia* in the Water Resources Act 1963, the Water Act 1973 and the Water Act 1989. It provides for the establishment and functions of the NRA and of committees to advise it; the transfer of property rights and liabilities of water authorities to the NRA; to make provisions to amend the law relating to the supply of water, the provision of sewers and the treatment and disposal of sewage, and the abstraction of water; to make new provision in relation to flood defence and fisheries; and to transfer functions with respect to navigation, conservancy and harbours to the NRA.

The land drainage functions of the NRA, work associated with main rivers, etc., are dealt with under this Act.

Land Drainage Act 1991

All the land drainage functions associated with other watercourses, IDBs, are dealt with under this Act.

Note: fisheries managers and angling club committees should note that this appendix provides the briefest of guides to some of the Acts that can affect the management of fisheries. The subject is complex and the 'amateur lawyer' should always check what is and what is not permissible under the law.

Appendix 8
Genetic developments and salmonid fishery management

In stillwater fisheries there is a small demand for interspecific fish crosses as a novelty. In practical terms, production of large numbers is not feasible due to high cost and low demand. However, the following crosses have been produced, and may be of interest to fish farmers.

Female rainbow × male brown = 'brownbow'
Female brownbow × male brown = 'sunbeam' (F_1 backcross)
Brook char × brown   = 'zebra' or 'tiger' (low fertility)
Brook char × lake char   = 'splake' (fertile)
Brook char × rainbow   = 'cheetah' (low fertility)
Brook char × Atlantic salmon (no popular name)

Of much greater use is the elimination of early sexual maturity in male rainbow trout which gives rise to the 'black rainbow': these fish are noted for their poor condition, unpleasant appearance and taste, and relatively poor growth rate. Elimination of undesirable sexual characteristics can be achieved in several ways, including: production of sterile fish; production of all female fish; radical delay in maturity, selective killing of male eggs and production of triploids.

Sterile fish can be produced by surgical castration or by administration of a high concentration of sex steroids. Other techniques are being developed, involving thermal shock to the eggs and irradiation of fish sperm. All-female fish can be produced successfully in salmonids by methods involving the use of hormones in the feed.

This whole topic has advanced considerably in recent years and the reader is referred to specialist textbooks and MAFF publications for a comprehensive treatment of the topic.

Appendix 9
Annual timetable of cyprinid farming

Below is an approximate timetable of the annual cycle of spawning, on growing and pond maintenance for UK cyprinid farming

January	Allow frost to break up silt structure. Repair 'monks' and pond banks. During prolonged freezing periods remove the ice from surface of wintering and brood fish ponds to allow gas transfer with the air.
February	Lime pond bottoms at least 3 to 4 weeks before filling commences. Check and fit screens, caulk weirs.
March	Begin to fill rearing ponds, manure when natural bloom commences. Cut grass in 'Dubisch' ponds, clear away dead vegetation.
April	Drain down brood fish. Grade and separate the sexes. Return to different holding ponds. Monitor development of zooplankton in rearing ponds. Keep daily check on pond temperatures.
May	Drain down overwintering ponds and stock C1+, C2+ and C3+ fish into rearing ponds with mature zooplankton bloom. Fill 'Dubisch' and nursery ponds when water temperature has reached 18–20°C. Spawn adults and stock larvae into nursery ponds. Check abundance of natural food in rearing and brood fish holding ponds. Commence regular manuring or supplementary feeding if necessary.
June	Increase feeding in rearing ponds. Commence feeding in nursery ponds, pump in fresh zooplankton if possible.
July	Continue feeding or manuring fish. Check growth rates. Monitor temperature, oxygen, pH and ammonia levels two or three times per week. If a problem develops, stop feeding or manuring and run in fresh water.
August	Same tasks as July. Ponds will now have high fish densities and must be watched carefully. Fish gasping at the surface is a sign of oxygen shortage. Overwintering ponds should now be cleaned out, filled with water and manured.
September	Water temperature begins to drop and supplementary feeding should be reduced. Overwintering ponds should now contain plenty of natural food.

October   Feeding ceases. Begin draining down ponds. Fish to be over-
              wintered are stocked into wintering ponds and large fish sold.
November   Finish harvesting ponds. Clear out all drainage ditches of
/December   accumulated silt and debris. Leave outlet valve open and allow
              the pond bottom to dry out. Fill in low points on pond bottom
              and remove excess mud and silt. Rake ponds to assist oxidation
              of the mud.

Appendix 10
Protected species list

Under the Wildlife and Countryside Act 1981 all wild birds, their nests and eggs are protected, with some exceptions for 'pests'. However, Schedule 1 of the Act gives special protection for the following birds:

Avocet	Bittern
Spotted crake	Diver (all species)
Long-tailed duck	Fieldfare
Garganey	Black-necked grebe
Slavonian grebe	Greenshank
Harriers (all species)	Hobby
Kingfisher	Red kite
Merlin	Barn owl
Peregrine falcon	Little ringed plover
Black redstart	Redwing
Ruff	Green sandpiper
Scaup	Common scoter
Bewick's swan	Whooper swan
Bearded tit	Marsh-warbler

Other animals:

Schedule 5 of the Act lists animals which are given special protection. The list is numerous and includes all whales. The following are most likely to be encountered:

Adder*	Bats (all)
Burbot	Swallowtail butterfly
Freshwater crayfish**	Dolphins (all)
Dormouse	Porpoises and whales
Common frog*	Medicinal leech
Common lizard***	Newts (Palmate*, Smooth* and Great Crested)
Otter	Slow-worm***
Grass snake***	Red squirrel
Common toad*	Marine turtles (all)

Vendace   Walrus
Whitefish

Recently added to the list:

Allis shad   Mire pill beetle
Pearl mussel   Sturgeon

*   = offence relates to sale of the species only
**   = offence relates to taking or sale of the species
***   = offence relates to killing, injury or sale

Appendix 11
Conservation designations of sites

Statutory sites protected in law

RAMSAR sites (convention on wetlands of international importance)
Areas of outstanding natural beauty (AONB)
National parks
Sites of special scientific interest (SSSI)
Environmentally sensitive areas (ESA)
Areas of special protection for birds
Special protection areas (EC Directive 79/409)
National nature reserve (NNR)
Local nature reserve (LNR)
Heritage coasts
Scheduled ancient monuments
Tree preservation orders
Conservation areas (planning)
Listed buildings

Non-statutory sites

Nature reserves
Site of importance for nature conservation (SINC)

Appendix 12
Conversion factors and useful equivalents

Weight

28.35 g	= 1 oz
0.4536 kg	= 1 lb
1 kg (1000 g)	= 2.205 lb (35.27 oz)
1 tonne	= 2205 lb
1.016 t	= 1 ton

Volume

4.546 litres	= 1 gallon
1 litre (1000 cm^3 or 1000 ml)	= 0.22 gallons
1 litre	= 1.76 pints (0.26 US gallons)
28.32 litres	= 1 ft^3
1 foot^3	= 6.23 gallons
0.7646 m^3	= 1 yard^3
1 metre^3	= 1.308 yard^3
1 m^3	= 35.31 ft^3

Length

25.4 mm (2.54 cm)	= 1 inch
30.48 cm	= 1 foot
0.914 m	= 1 yard
1 metre	= 39.37 inches
1.6093 km	= 1 mile

Area

0.836 m^2	= 1 yard^2
1 m^2	= 1.196 yard^2
4840 yard^2	= 1 acre
2.47 acres	= 1 hectare

Water equivalent

10 lb (4.54 kg)	= 1 gallon
6.23 gallons (28.3 kg)	= 1 ft^3

Flow rate

1.26 ml per second	= 1 gallon per hour
1 litre per second	= 13.2 gallons per minute
28.32 litre per second (373.8 gpm)	= 1 ft^3 per second
538 272 gallons per day	= 1 ft^3 per second
4.546 megalitres	= 1 million gallons
1 m^3	= 220.083 gallons

Temperature

On the centigrade scale 0°C and 100°C represent the freezing and boiling points respectively of water (at standard pressure). To convert Fahrenheit (°F) degrees to Centigrade (°C) use the formula:

$$°F = (°C \times 9/5) + 32$$

Appendix 13
Common animal and plant names, with scientific equivalents

Birds

Black-headed gull	– *Larus ridibundus* (L)
Canada goose	– *Branta canadensis* (L)
Cormorant	– *Phalacrocorax carbo* (L)
Great black-backed gull	– *Larus marinus* (L)
Great crested grebe	– *Podiceps cristatus* (L)
Grey heron	– *Ardea cinerea* (L)
Herring gull	– *Larus argentatus* (L)
Kingfisher	– *Alcedo atthis* (L)
Lesser black-backed gull	– *Larus fuscus* (L)
Little grebe (dabchick)	– *Podiceps ruficollis* (L)
Tufted duck	– *Aythya fuligula* (L)

Mammals

Coypu	– *Myocaster coypus* (Molina)
Mink	– *Mustela vison* (L)
Mole	– *Talpa europaea* (L)
Otter	– *Lutra lutra* (L)
Rat, common	– *Rattus norvegicus* (Berkenhart)
Water vole	– *Arvicola terrestris* (L)

Fish

Barbel	– *Barbus barbus* (L)
Bitterling	– *Rhodeus sericeus* (Bloch)
Bleak	– *Alburnus alburnus* (L)
Bream	– *Abramis brama* (L)
Bullhead	– *Cottus gobio* (L)
Carp	– *Cyprinus carpio* (L)
Carp, chinese grass	– *Ctenopharyngoden idella* (Val)
Carp, crucian	– *Carassius carassius* (L)
Char (Arctic)	– *Salvelinus alpinus* (L)
Chub	– *Leuciscus cephalus* (L)
Dab	– *Limanda limanda* (L)
Dace	– *Leuciscus leuciscus* (L)

Eel	– *Anguilla anguilla* (L)
Flounder	– *Platichthys flesus* (L)
Goldfish	– *Carassius auratus* (L)
Grayling	– *Thymallus thymallus* (L)
Gudgeon	– *Gobio gobio* (L)
Lamprey, brook	– *Lampetra planeri* (Bloch)
Lamprey, river	– *Lampetra fluviatilis* (L)
Loach stone	– *Noemacheilus barbatulus* (L)
Minnow	– *Phoxinus phoxinus* (L)
Mullet, thick-lipped	– *Chelon labrosus labrosus* (Risso)
Mullet, thin-lipped	– *Liza ramada* (Risso)
Perch	– *Perca fluviatilis* (L)
Pike	– *Esox lucius* (L)
Roach	– *Rutilus rutilus* (L)
Rudd	– *Scardinius erythrophthalmus* (L)
Ruffe	– *Gymnocephalus cernua* (L)
Salmon	– *Salmo salar* (L)
Silver bream	– *Blicca bjoerknall* (L)
Stickleback, 3 spined	– *Gasterosteus aculeatus* (L)
Tench	– *Tinca tinca* (L)
Trout, American brook (Brook char)	– *Salvelinus fontinalis* (Mitchill)
Trout, brown	– *Salmo trutta* (L)
Trout, rainbow	– *Onchorhynchus mykiss* (Walbaum)
Trout, sea	– *Salmo trutta trutta* (L)
Zander	– *Stizostedion lucioperca* (L)

Invertebrates

Beetle larvae	– *Coleptera*
Caddis larvae	– *Trichoptera*
Freshwater hog louse	– *Ascellus aquaticus*
Freshwater shrimp	– *Gammarus* spp
Pea mussels	– *Pisidium* spp
Swan mussels	– *Anodonta* spp

Parasites

Eye fluke	– *Diplostomum spathaceum* (L)
Fish louse	– *Argulus* spp
Leech	– *Piscicola* spp
Tapeworm-pike	– *Triaenophorus nodulosus* (Pallus)
Tapeworm	– *Diphylobothrium* spp
Tapeworm	– *Ligula* spp
Whirling disease	– *Myxosoma cerebralis* (Hofer)

Plants (see also Appendix 6)

Amphibious bistort	– *Polygonum amphibium* (L)
Arrowhead	– *Sagittaria sagittifolia* (L)
Bullrush	– *Scirpus lacustris* (L)
Bur-reed	– *Sparganium erectum* (L)
Canadian pond weed	– *Elodea canadensis* (Michx)
Common reed	– *Phragmites communis* (Trin)
Duckweed	– *Lemna minor* (L)
Frogbit	– *Hydrocharis morsus-ranae* (L)
Great water dock	– *Rumex hydrolapathum* (Huds)
Hornwort	– *Ceratophyllum demersum* (L)
Ivy-leaved duckweed	– *Lemna trisulca* (L)
Mare's tail	– *Hippuris vulgaris* (L)
Pondweed	– *Potomogeton* spp
Reed grass	– *Glyceria maxima* (Hartm)
Reed mace	– *Typha latifolia* (L)
Stonewort	– *Chara* spp
Watercress	– *Nasturtium officinale* (R. Br)
Water crowfoot	– *Ranunculus aquatilis* (L)
Water lilies	– *Nymphaea alba* (L)
	– *Nuphar lutea* (L)
Water milfoil	– *Myriophyllum* spp
Water plantain	– *Alisma plantago aquatica* (L)
Water starwort	– *Callitriche stagnalis* (Scop)

Trees

Alder	– *Alnus glutinosa* (L)
Ash	– *Fraxinus excelsior* (L)
Crack willow	– *Salix fragilis* (L)
Hairy birch	– *Betula pubescens* (Ehrh)
Oak	– *Quercus robur* (L)
White willow	– *Salix alba* (L)

Reading list

Part 1 The resource

Alabaster J.S. (1990) Water quality for freshwater fish – review of progress. *IFM Annual Study Course Proc.*

Bagenal T.B. (1973) *Identification of British Fishes*. Educational Publishers.

Bagenal T.B. (1978) *Methods for Assessment of Fish Production in Freshwaters*. IBP Handbook No. 3. Blackwell Science, Oxford.

Bobek M. (1990) Applied hydroacoustics in cyprinid research. *IFM Annual Study Course Proc.*

Cooper M.J. (1979) Large river monitoring. *Proc. of IFM Annual Study Course.*

Cowx I. & Lararque (1991) *Fishing with Electricity*. Fishing News Books, Oxford.

Currie K. (1992) The early history of the carp and its economic significance in England. *FISH No. 26. IFM.*

Dunkley D.A. (1991) The use of fish counters in the management of salmonid stocks: the example of the North Esk. *IFM Annual Study Course Proc.*

Englehardt W. (1973) *Pond Life*. Burke Publishing.

Harris J.R. (1973) *An Angler's Entomology*. New Naturalist, Collins.

Hofferman G. & Meyer F.P. (1974) *Parasites of Freshwater Fishes*. TFH Publishers.

Howarth W. (1988) *Water Pollution Law, with Supplements*. Shaw & Sons Ltd.

Hynes H.B.N. (1972) *Ecology of Running Waters*. Liverpool University Press.

Jeffries & Mills D. (1990) *Freshwater Ecology*. Bellhaven Press.

Jones R. (1979) Materials and methods used in marking experiments in fishery research. *FAO Fisheries Technical Rep. No. 190.*

Kennedy G.J.A. & Strange C.D. (1981) Efficiency of electric fishing for salmonids in relation to river width. *Fisheries Management.* No. 12(2) 55–60.

Lloyd R. (1992) *Pollution and Freshwater Fish*. Fishing News Books.

Macan T.T. & Worthington E.D. (1972) *Life In Lakes and Rivers*. New Naturalist, Fontana.

Maclennan D.N. & Simmonds E.J. (1992) *Fisheries Acoustics*. Chapman and Hall.

Maitland P.S. (1972) Key to British Freshwater Fishes. *FBA Scientific Publication No 27.*

Maitland P.S. & Campbell R.N. (1992) Freshwater Fishes. New Naturalist, Harper Collins.

Mills D. (1971) *An Introduction to Freshwater Ecology*. Oliver & Boyd.

Mills D. (1971) *Salmon and Trout: A Resource, its Ecology, Conservation and Management.*

Mills D. (1989) *Ecology and Management of Atlantic Salmon.* Chapman and Hall, London.

Mitson R.B. (1983) *Fisheries Sonar.* Fishing News Books.

Moriarty F. (1983) *Ecotoxicology. The Study of Pollutants in Ecosystems.* 2nd edition. Academic Press.

Moss B. (1980) *Ecology of Freshwaters.* Blackwell Science, Oxford.

Perrow M.R. (1991) Reversing the effects of eutrophication upon fish communities: lessons from Broadland. *IFM Annual Study Course Proc.*

Pickering A.D. (1982) *Stress and Fish.* Academic Press.

Pratt M.M. (1975) *Better Angling with Simple Science.* Fishing News Books.

Priede I. G. (1991) Telemetry in assessment of environmental effects on individual fishes. *IFM Annual Study Course Proc.*

Seymour, R. (1970) *Fisheries Management and Keepering.* Charles Knight and Co. Ltd.

Smith G.W. (1991) Salmon movements in relation to river flow: estuarine net catches of adult Atlantic salmon and tracking observations. *IFM Annual Study Course Proc.*

Towner Coston *et al.* (1986) *River Management.*

Varley M.E. (1967) *British Freshwater Fishes.* Fishing News Books.

Watson D.C. (1988) *Fisheries Acidification.* IFM Booklet.

Welton J.S., Beaumont W.R.C. & Johnson I.K. (1988) Salmon counting in chalk-streams. *IFM Annual Study Course Proc.*

Wheeler A. (1978) *Key to the Fishes of Northern Europe.* F. Warne.

Part 2: Management

Angling Foundation (The) (1974) *The Creation of Low-Cost Fisheries.*

Birch E. (1964) *The Management of Coarse Fishing Waters.* J. Baker: London.

Bursche E.M. (1971) *A Handbook of Water Plants.* F. Warne.

Cowx I. G. & Lamarque P. (1990) *Fishing with Electricity: Applications in Freshwater Fisheries Management.* Fishing News Books.

Cryer M. & Edwards R. (1987) The impact of angler ground bait on benthic invertebrates and sediment respiration in a shallow eutrophic reservoir. *Environmental Pollution* 46: 137–150.

Cuthbert J.H. (1979) Food studies of feral mink. *Fisheries Management* No. **10**(1): 17–26.

Feare C.J. (1987) Cormorants as predators at freshwater fisheries. *IFM Annual Study Course Proc.*

Frost W. (1967) *The Trout.* New Naturalist, Collins.

Haslam S., Sinker C. & Wolsely (1975) *British Water Plants.* Field Studies Council, Shrewsbury.

Hey, R.D. (1990) River mechanics and habitat creation. *IFM Annual Study Course Proc.*

Hopkins T. & Brassley (1982) *Wildlife of Rivers and Canals.* Moorlands Publishing, Ashbourne, Derbys.

Howarth W. (1987) *Freshwater Fishery Law.* Blackstone Press.

Howarth W. (1992) *Wisdom's Law of Watercourses.* Shaw & Sons Ltd.

Keble Martin W. (1969) *The Concise British Flora in colour.* Ebury Press, M. Joseph.

Marton R.K. (1971) *Man and bird.* New Naturalist, Collins.

Mills D. (1979) Bird predation – current views. *IFM Annual Study Course Proc.*

Morrison B.R.S. The influence of forestry management on freshwater fisheries with particular reference to salmon and trout. *IFM Annual Study Course Proc.*

RSPCA (1981) *Book of Mammals.* Collins.

Seagrave C. (1988) *Aquatic Weed Control.* Fishing News Books.

Wolds A. *et al.* (1992) Effect of ground-bait on catches and nutrient budget. *Aquaculture and Fisheries Management* **23**(4): pp. 499–510.

Part 3: Exploitation

Abrahamsson Sture (1973) Freshwater crayfish. *Papers from the First International Symposium on Freshwater Crayfish,* Austria 1972.

Bardach J. Ryther J. Clarney W.M. (1977) *Aquaculture.* Wiley Interscience, 2nd Ed.

Barnabe G. (1990) *Aquaculture,* Vols 1 and 2. Ellis Horwood Ltd.

Brown E. & Gratzek J. (1980) *Fish Farming Handbook.* Van Nostrand Reinhold.

Burgess J. *Trammel Netting.* Bridport-Gundry Ltd, Dorset.

Burgess J. *Drift, Gill and Ray netting.* Bridport-Gundry Ltd, Dorset.

Davies H.S. (1961) *Culture and Diseases of Game Fishes.* University of California Press and Cambridge University Press, London.

Dear G. Practical aspects of fish disease treatment. *IFM Annual Study Course Proceedings.*

Drummond-Sedgwick S. (1990) *Trout Farming Handbook* (5th Ed.) Fishing News Books, Oxford.

Edwards R. (1990) The impact of angling on conservation and water quality. *IFM Annual Study Course Proceedings.*

Forrest D.M. (1976) *Eel Capture, Culture, Processing and Marketing.* Fishing News Books.

Garner J. (1962) *How to Make and Set Nets.* Fishing News Books.

Greenberg D.G. (1980) *Trout Farming.* Chilton Co., Philadelphia.

Hand M. (1992) Predation in Fish Farms (a personal view) *Fish* No 27, IFM.

Hepher B. & Pruginin Y. (1981) *Commercial Fish Farming.* Wiley Interscience.

Holditch D.M. Lowery S. (1988) *Freshwater Crayfish, Biology, Management and Exploitation.*

Horvath L. Tamas G. & Seagrave C. (1992) *Carp and Pondfish Culture.* Fishing News Books.

Horvath L. Tamas G. & Tolg I. (1984) *Special Methods in Pond Fish Husbandry.* Halver Corporation.

Howarth W. (1990) *The Law of Aquaculture.* Fishing News Books.

Hudson E.B. (1990) Developments in fish disease control. *IFM Annual Study Course Proceedings.*

Huet M. & Timmermans J. (1986) *Textbook of Fish Culture* (2nd Ed.) Fishing News Books.

IFM Certificate booklet on *fishing methods.*

IFM Certificate and Diploma booklets on *fish propagation*.

Inglis V. Roberts R. J. & Bromage N. (1993) *Bacterial Diseases of Fish*. Blackwell Science, Oxford.

Jhingran V.G. & Pullin R.S.V. (1988) *A Hatchery Manual for the Common Carps*. Asian Dev. Bank/ICLARM, 2nd Ed.

Karlssom S. (1977) The freshwater crayfish. *Fish Farmer Magazine*.

Leitritz E. and Lewis C. (1980) *Trout and Salmon Culture* (Hatchery Methods). Fish Bulletin No. 164 of the State of California Department of Fish and Game. Publication 4100 Agricultural Science Publications, University of California.

Michaels V.K. (1988) *Carp Farming*. Fishing News Books.

Mills D. (1971) *Salmon and Trout: A Resource, Its Ecology, Conservation and Management*. Oliver and Boyd, Edinburgh.

Netboy A. (1968) *The Atlantic Salmon. A Vanishing Species?* Faber and Faber, London.

Pillay T.V.R. (1990) *Aquaculture: Principles and Practices*. Fishing News Books.

Piper, McElwain, Orme *et al.* (1982) *Fish Hatchery Management*. USFWS.

Post G. (1987) *Textbook of Fish Health*. T.F.H. Publications, 2nd Ed.

Reay P. (1979) *Aquaculture*. Institute of Biology Publication 106.

Reynolds J.D. (1989) A perspective on crayfish culture in the British Isles. *IFM Annual Study Course Proceedings*.

Roberts R.J. (1989) *Fish Pathology*. Balliere Tindall, 2nd Ed.

Scottish District Salmon Fishery Boards, Association of (1977) *Salmon Fisheries of Scotland*. Fishing News Books.

Sinha V.R.P. & Jones J.W. (1975) *The European Freshwater Eel*. Liverpool University Press.

Stevenson J.P. (1980) *Trout Farming Manual*. Fishing News Books.

Stickney R. (1979) *Principles of Warm Water Aquaculture*. Wiley.

Tesch F.W. (1977) *The Eel Biology and Management of Anguillid Eels*. Chapman and Hall, London.

Untergasser D. (1989) *Handbook of Fish Diseases*. T.F.H. Publications.

Usui A. (1991) *Eel Culture*. Fishing News Books, 2nd Ed.

Woynarovich E. & Horvath L. (1980) *The Artificial Propagation of Warm Water Finfishes*. F.A.O.

Note: Many of these publications can be purchased from the Institute of Fisheries Management's Mail Order Catalogue – see Appendix 3 for address.

The Atlantic Salmon Trust produce a series of booklets on various aspects of salmonid management.

Index

Books published by
Fishing News Books

Free catalogue available on request from Fishing News Books, Blackwell Scientific Publications Ltd, Osney Mead, Oxford OX2 0EL, England

Abalone farming
Abalone of the world
Advances in fish science and technology
Aquaculture and the environment
Aquaculture & water resources management
Aquaculture development – progress and
  prospects
Aquaculture: principles and practices
Aquaculture in Taiwan
Aquaculture systems
Aquaculture training manual
Aquatic ecology
Aquatic microbiology
Aquatic weed control
Atlantic salmon: its future
The Atlantic salmon: natural history etc.
Bacterial diseases of fish
Better angling with simple science
Bioeconomic analysis of fisheries
British freshwater fishes
Broodstock management and egg and larval
  quality
Business management in fisheries and
  aquaculture
Cage aquaculture
Calculations for fishing gear designs
Carp farming
Carp and pond fish culture
Catch effort sampling strategies
Commercial fishing methods
Common fisheries policy
Control of fish quality
Crab and lobster fishing
The crayfish
Crustacean farming
Culture of bivalve molluscs
Design of small fishing vessels
Developments in electric fishing
Developments in fisheries research in Scotland
Dynamics of marine ecosystems
Ecology of fresh waters
The economics of salmon aquaculture
The edible crab and its fishery in British waters
Eel culture
Engineering, economics and fisheries
  management
The European fishing handbook 1993–94
FAO catalogue of fishing gear designs
FAO catalogue of small scale fishing gear
Fibre ropes for fishing gear
Fish catching methods of the world
Fisheries biology, assessment and management
Fisheries oceanography and ecology
Fisheries of Australia
Fisheries sonar
Fishermen's handbook
Fisherman's workbook
Fishery development experiences
Fishery products and processing
Fishing and stock fluctuations
Fishing boats and their equipment
Fishing boats of the world 1
Fishing boats of the world 2
Fishing boats of the world 3
Fishing ports and markets
Fishing with electricity
Fishing with light
Freshwater fisheries management
Fundamentals of aquatic ecology
Glossary of UK fishing gear terms

Handbook of trout and salmon diseases
A history of marine fish culture in Europe and
  North America
How to make and set nets
The Icelandic fisheries
Inland aquaculture development handbook
Intensive fish farming
Introduction to fishery by-products
The law of aquaculture: the law relating to the
  farming of fish and shellfish in Great Britain
A living from lobsters
Longline fishing
Making and managing a trout lake
Managerial effectiveness in fisheries and
  aquaculture
Marine climate, weather and fisheries
Marine fish behaviour in capture and abundance
  estimation
Marine fisheries ecosystem
Marketing: a practical guide for fish farmers
Marketing in fisheries and aquaculture
Mending of fishing nets
Modern deep sea trawling gear
More Scottish fishing craft and their work
Multilingual dictionary of fish and fish products
Multilingual dictionary of fishing vessels/safety on
  board
Multilingual dictionary of fishing gear
Multilingual illustrated dictionary of aquatic
  animals & plants
Navigation primer for fishermen
Netting materials for fishing gear
Net work exercises
Ocean forum
Pair trawling and pair seining
Pelagic and semi-pelagic trawling gear
Pelagic fish: the resource and its exploitation
Penaeid shrimps — their biology and management
Planning of aquaculture development
Pollution and freshwater fish
Purse seining manual
Recent advances in aquaculture IV
Recent advances in aquaculture V
Refrigeration on fishing vessels
Rehabilitation of freshwater fisheries
The rivers handbook, volume 1
The rivers handbook, volume 2
Salmon and trout farming in Norway
Salmon aquaculture
Salmon farming handbook
Salmon in the sea/new enhancement strategies
Scallop and queen fisheries in the British Isles
Scallop farming
Seafood science and technology
Seine fishing
Shrimp capture and culture fisheries of the US
Spiny lobster management
Squid jigging from small boats
Stability and trim of fishing vessels and other
  small ships
The state of the marine environment
Stock assessment in inland fisheries
Study of the sea
Sublethal and chronic toxic effects of pollution on
  freshwater fish
Textbook of fish culture
Trends in fish utilization
Trends in ichthyology
Trout farming handbook
Tuna fishing with pole and line